THE EUROPEAN UNION AND ASEAN:
TRADE AND INVESTMENT ISSUES

The European Union and ASEAN: Trade and Investment Issues

Edited by

Roger Strange, Jim Slater and
Corrado Molteni

First published in Great Britain 2000 by
MACMILLAN PRESS LTD
Houndmills, Basingstoke, Hampshire RG21 6XS and London
Companies and representatives throughout the world

A catalogue record for this book is available from the British Library

ISBN 0–333–77506–6

First published in the United States of America 2000 by
ST. MARTIN'S PRESS, LLC,
Scholarly and Reference Division,
175 Fifth Avenue, New York, N.Y. 10010

ISBN 0–312–23184–9

Library of Congress Cataloging-in-Publication Data
The European Union and ASEAN: trade and investment issues / edited by
Roger Strange, Jim Slater and Corrado Molteni.
p.   cm.
Includes bibliographical references and index.
ISBN 0–312–23184–9 (cloth)
1. European Union countries—Commerce—Asia, Southeastern. 2. Asia,
Southeastern—Commerce—European Union countries. 3. Asia,
Southeastern—Economic conditions. 4. Investments, European—Asia,
Southeastern. 5. Investments, Asian—European Union countries.
6. ASEAN. I. Strange, Roger. II. Slater, Jim, 1947–
III. Molteni, Corrado.

HF3498.A85 E96   2000
382′.094059—dc21   99-056732

This book is printed on paper suitable for recycling and made from fully managed and
sustained forest sources.

10   9   8   7   6   5   4   3   2   1
09   08   07   06   05   04   03   02   01   00

Printed in Great Britain by
Antony Rowe Ltd, Chippenham, Wiltshire

Contents

Contents vii

List of Figures

ix

List of Tables

Abbreviations

AD	anti-dumping
AFTA	ASEAN Free Trade Area
AIA	ASEAN Investment Area
APEC	Asia–Pacific Economic Cooperation
ASEAN	Association of Southeast Asian Nations
ASEAN-4	Indonesia, Malaysia, the Philippines, and Thailand
ASEAN-5	ASEAN-4, plus Singapore
ASEM	Asia–Europe Meeting
BIBF	Bangkok International Banking Facilities
BIS	Bank for International Settlements
BITS	bilateral investment treaties
BOT	build-operate-transfer (companies)
CAP	Common Agricultural Policy (of the European Union)
CEE	Central and Eastern European (countries)
CEPR	Centre for Economic Policy Research
CLIO	Conversational Language for Input–Output
CN	Combined Nomenclature of the European Communities
CNCT	Conseil National du Crédit et du Titre
DOS	Department of Statistics (Malaysia)
EAP	East Asia and the Pacific (countries)
EC	European Community
ecu	European Currency Unit
EDB	Economic Development Board (Singapore)
EIU	Economist Intelligence Unit
EMS	European Monetary System
EMU	European Monetary Union
EPZ	export processing zone
ERM	Exchange Rate Mechanism (of the European Union)
EU	European Union
EU15	The 15 Member States of the European Union (1999)
EU25	The 25 Potential Member States of the European Union
EXIM	Export–Import Bank of Japan
FDI	foreign direct investment
FIL	Foreign Investment Law (of Vietnam)

xiii

GDP	Gross Domestic Product
GL	Grubel–Lloyd index of intra-industry trade
GNP	Gross National Product
GSP	Generalized System of Preferences
HIIT	horizontally-differentiated intra-industry trade
HPAE	high-performing Asian economies
HS	Harmonized Commodity Description and Coding System
IIT	intra-industry trade
IMF	International Monetary Fund
IZ	industrial zone
JOI	Japan Institute for Overseas Investment
MFA	Multi-Fibre Arrangement
MFN	most favoured nation
MITI	Ministry of International Trade and Industry (Malaysia)
NACE	Nomenclature Générale des Activités Economiques dans les Communautés Européenes
NBER	National Bureau of Economic Research
NIEs	Newly Industrializing Economies
ODI	outward direct investment
OECD	Organization for Economic Co-operation and Development
PATA	Pacific Asia Travel Association
PLC	product life cycle (hypothesis)
PNG	Papua New Guinea
POCPA	Palm Oil Credit and Payments Agreement (Malaysia)
PRC	People's Republic of China
R&D	research and development
RCA	revealed comparative advantage
REI	regional economic integration
RM	Malaysian ringgit
SEA-10	Ten countries in Southeast Asia (i.e. Indonesia, Malaysia, the Philippines, Singapore, Thailand, Brunei, Vietnam, Laos, Myanmar and Cambodia)
SEM	Single European Market
SMEs	small and medium-sized enterprises
TAT	Tourism Authority of Thailand
TNCs	transnational corporations
UK	United Kingdom

UNCTAD	United Nations Conference on Trade and Development
US	United States (of America)
VER	voluntary export restraints
VHIIT	vertically-differentiated intra-industry trade, where the unit value of exports is higher than the unit value of imports
VIIT	vertically-differentiated intra-industry trade
VLIIT	vertically-differentiated intra-industry trade, where the unit value of exports is lower than the unit value of imports
WTO	World Trade Organization
WTTC	World Travel and Tourism Council

List of Contributors

Bernadette Andréosso-O'Callaghan is Jean Monnet Professor of European Economic Integration and Director of the Euro-Asia Centre, University of Limerick. Her major research interests lie in the area of economic integration, with a particular focus on EU–Asian relations, and on structural change in European and Asian countries. Her publications include *Industrial Economics and Organization – a European Perspective* (McGraw-Hill, 1996, with David Jacobson), and *Changing Economic Environment in Asia – Implications for Firms' Strategies in the Region* (Macmillan, 2000, co-edited with Jean Pascal Bassino and Jacques Jaussaud).

Frank Bartels is Assistant Professor in International Business at Nanyang Technological University. His recent publications include 'Towards a Strategy for Enhancing ASEAN's Locational Advantages for Attracting Greater Foreign Direct Investment' (1998, with Hafiz Mirza and Kee Hwee Wee) and 'Multinational Corporations and Foreign Direct Investment in Asia's Emerging Markets: Before and After the Economic Crisis: Any Changes?' (1999, with Hafiz Mirza).

Anthony Bende-Nabende is a Research Associate in the Business School at the University of Birmingham. His research interests are focused on foreign direct investment, regionalism and growth theories, particularly with regard to their policy implications for developing countries. He is the author of *FDI, Regionalism, Government Policy and Endogenous Growth* (Avebury, 1999).

Axèle Giroud is Lecturer in International Business and Chair of the MA in International Business and Management at the University of Bradford Management Centre. Her publications include *The Promotion of Foreign Direct Investment into and within ASEAN: Towards the Establishment of an ASEAN Investment Area* (ASEAN Secretariat, 1997, with Hafiz Mirza, Frank Bartels and Mark Hiley) and 'Upgrading Endogenous Production Capabilities in LDCs: an Evaluation of Information Flows between TNC Subsidiaries and their Local Suppliers' (1999).

Robert Hine is Senior Lecturer in the School of Economics at the University of Nottingham, and Visiting Professor at the College of Europe, Bruges. His teaching and research interests are in the field of European trade and economic integration, and his recent publications include *Intra-Industry Trade and Adjustment* (Macmillan, 1999, with Brulhart) and *The Reform of the Common Agricultural Policy* (Macmillan, 1998, with Rayner and Ingersent).

Diana Hochraich is an economist and an expert on the economics of development, particularly with regard to the Asian countries. She has recently published *L'Asie, du Miracle à la Crise* (Complexe, 1999) and *La Chine de la Revolution à la Reforme* (Syros, 1995).

John Joyce is a postgraduate researcher in the Euro-Asia Centre at the University of Limerick. His research is in the area of economic transformation in Vietnam.

Hafiz Mirza is Professor of International Business and Chair of the Asia–Pacific Business and Development Research Unit (APBDRU) at the University of Bradford Management Centre. He has published widely, and recent books include *Global Competitive Strategies in the New World Economy: Multilateralism, Regionalisation and the Transnational Firm* (Edward Elgar, 1998) and *International Strategic Management* (Macmillan, 1998, with Peter J. Buckley and Fred Burton).

Corrado Molteni is Associate Professor of Japanese Studies in the Faculty of Political Science at Milan State University, Adjunct Professor of Comparative Economic Systems at Bocconi University, Milan, and Coordinator of the European Studies Programme at the University of Malaya, Kuala Lumpur, Malaysia. He has published widely, in both English and Italian, on the East Asian economies.

Jim Newton teaches International Business at the University of Hong Kong's School of Business and was previously Director of the University's MBA Programme. His research interests concern issues in international trade in both goods and services, especially in Asia and between Asia and the West. The focus of current studies is on the international political economy of specific sectors of importance to

Hong Kong, China and Asia. He is also a contributor to the School's Centre for Asian Business Cases which is producing a library of case studies for use in both local and international management and business programmes.

Françoise Nicolas is a Senior Research Fellow at the French Institute of International Relations, Paris and an Associate Lecturer in International Economics at the Université de Marne-la-Vallée. She holds a Ph.D. in International Economics from the Graduate Institute of International Studies (GIIS) in Geneva, and a Licence in Political Science from GIIS and the University of Geneva. She has served as a consultant to the OECD and has recently published *Foreign Direct Investment and Recovery in Southeast Asia* (OECD, 1999, with Stephen Thomsen).

Richard Pomfret has been Professor of Economics at the University of Adelaide since 1992. From 1979 to 1991 he was at the Johns Hopkins School of Advanced International Studies, and taught in Baltimore and Washington (USA), Bologna (Italy) and Nanjing (China). In 1993 he spent a year with the United Nations Economic and Social Commission for Asia and the Pacific, and he has worked as a consultant for the UNDP, the World Bank and other international agencies. His major fields of research are economic development and international economics. He has written fourteen books, most recently *The Economics of Regional Trading Arrangements* (Oxford University Press, 1997).

Jim Slater is Director of the Graduate Centre for Business Administration at the University of Birmingham and has, since 1985, been responsible for the development of the MBA programme. He has extensive research and management education links in Europe, South East Asia, Africa and Latin America, and his consultancy experience includes a number of private and public institutions. His present research is focused on foreign direct investment, and he is the co-editor (with Roger Strange) of *Business Relationships with East Asia: the European Experience* (Routledge, 1997) and *Trade and Investment in China: the European Experience* (Routledge, 1998).

Roger Strange is Senior Lecturer in Economics, and formerly Head of the Management Centre, at King's College London. He has published widely on Europe–Asia economic relations. His most recent books are *Japanese Manufacturing Investment in Europe: its Impact on the UK Economy* (Routledge, 1993), *Business Relationships with East Asia* (Routledge, 1997, edited with Jim Slater), *Management in China: the Experience of Foreign Businesses* (Frank Cass, 1998, editor) and *Trade and Investment in China* (Routledge, 1998, edited with Jim Slater and Limin Wang).

Isabel Tirado Angel is a project analyst with the global consulting group, W.S. Atkins. She took her first degree in Civil Engineering at the Universidad de los Andes in Colombia, and then practised for three years as an engineer before studying for an MBA at Birmingham University. She then read for her Ph.D. at Birmingham University, specialising in the implications of ASEAN foreign direct investment, before taking up her current post.

Kee Hwee Wee is Assistant Director, Services, Investment and Finance at the ASEAN Secretariat and is currently overseeing the establishment of the ASEAN Investment Area (AIA) and Regional Financial Surveillance across South East Asia. His publications include *The Strategy, Objectives and Performance of Transnational Corporations in ASEAN: Country Studies* (Edward Elgar, 2000, with Hafiz Mirza).

Introduction
Roger Strange

In mid-1997, a major crisis embraced the economies of Southeast Asia. Its effects were severe, particularly on the countries of the Association of Southeast Asian Nations (ASEAN),[1] many of whose currencies suffered marked devaluations. Prior to the crisis, many of these countries had been experiencing prolonged periods of economic growth: indeed four of them[2] had been identified as 'high-performing Asian economies' (HPAEs) by the World Bank in its assessment of the 'East Asian Miracle'.

Europe, and in particular the European Union (EU),[3] was starting to show increasing interest. In 1994, the European Commission published a strategy document calling for stronger ties with the Asian economies (Commission of the European Communities 1994). The Asia–Europe Meetings (ASEM) were introduced: the first of which took place in Bangkok in 1996, and the second in London in 1998. The ASEAN markets were becoming increasingly important destinations for EU exports, and more important locations for EU foreign direct investment (FDI).

The crisis changed both the reality of EU–ASEAN economic relations and everyone's perceptions. Previously, the HPAEs had been held up as models for economic development and their flaws had been carefully overlooked. Businessmen had flocked to the region in search of profitable opportunities, and academics had waxed lyrical about the reasons for their success. Since then, the deficiencies have been highlighted and the success story of the 1980s and 1990s has been downplayed.

Where does this leave EU–ASEAN economic relations at the turn of the century? This is the central question addressed by the various chapters in this edited volume. Many of the chapters were first presented at a conference on EU–ASEAN relations which was organized by the Universita Commerciale Luigi Bocconi and held at Como, Italy in April 1998 (i.e. almost one year after the crisis erupted, and at about the same time as the Second ASEM

1

conference in London). Other chapters were commissioned by the editors.

The chapters are grouped in four sections. The first section looks at the macroeconomic picture, and examines trade and investment flows in both directions between the European Union and ASEAN. In Chapter 1, Robert Hine presents a detailed analysis of EU–ASEAN trade flows. He points out the asymmetry in the trade relationship with the European Union being a much more important trade partner for the ASEAN countries than vice versa. Furthermore, although he finds evidence of a general convergence in EU–ASEAN trade with intra-industry trade becoming ever more important than inter-industry trade, he also identifies significant quality differences. He concludes that the EU–ASEAN bridge has an important role to play in the recovery of the Asian economies from the 1997 crisis, but that this may generate significant adjustment pressures and protectionist demands in Europe.

In Chapter 2, Anthony Bende-Nabende takes an historical look at FDI in the five largest ASEAN economies (ASEAN-5) in an attempt to explain the declining EU share of total inward investment. He catalogues the shifts in the ASEAN industrialization strategies from the emphasis on primary production during the colonial era, to import-substitution in the early post-independence period, and thence to export-oriented development. He argues that the changes in development strategy were accompanied by production diversification towards non-traditional manufacturing activities, and by liberalization of the regimes regarding inward investment. Furthermore these changes took place against the backdrop of a realignment of the major currencies in the mid-1980s, and that companies from Japan and the Asian Newly Industrializing Economies (NIEs) were quick to use FDI to rationalize their activities throughout Asia. EU companies meanwhile were more concerned with the coming of the Single European Market (SEM) and with developments in central and Eastern Europe.

Hafiz Mirza *et al.* report in Chapter 3 the results of a major survey of foreign firms' investment intentions with regard to ASEAN, carried out just before the Asian crisis. They too note that EU FDI in ASEAN began to falter in the early 1990s, and suggest that the ASEAN countries should endeavour to reverse this trend. They emphasize the need for the ASEAN countries to grow together in terms of the ASEAN Free Trade Area (AFTA) and the ASEAN Investment Area. Encouragingly, they find no evidence of sudden post-crisis disinvestment by foreign firms from ASEAN, and are optimistic about the prospects for future FDI in the region. In contrast to EU FDI in the ASEAN countries, ASEAN FDI in the European Union is much smaller both in absolute and in relative terms, though the total amount of outward direct investment from Southeast Asia was on the increase prior to the 1997 crisis. Even this statement, however, needs to be qualified by the fact that reliable statistics are not available. In Chapter 4, Jim Slater outlines the ASEAN governments' policies on outward direct investment (ODI), and highlights the incentives and political support provided by the Malaysian Government in particular. He also pulls together data from a variety of sources on ASEAN FDI in Europe, but concludes that most is in financial services and that there is very little in manufacturing industry.

The second section considers the reasons for the crisis of 1997, and its repercussions both for the countries directly involved and for Europe. Françoise Nicolas compares the Asian crisis of 1997 with the 1992–93 crisis in the European Monetary System (EMS). She notes that the crises were different in many respects, but that the EMS experience does shed light on the process of contagion of the Asian crisis across countries in the region. She also considers the impact of the Asian crisis on Europe, both through its effects on trade and investment, and through the financial channel. She concludes that the direct effects on trade and investment are likely to be negligible in the short and medium-term, but that the indirect effects (in particular via the slowdown in US economic growth) may be more substantial. In contrast, she argues that the financial effects have been rather more important as European banks had a high exposure to Asian risk prior to the crisis, and that they have subsequently had to increase their loan provisions. She concludes by noting the surprisingly rather limited role played by the European Union in the various rescue packages for the Asian economies, and suggests that the crisis has provided the EU Member States with an opportunity to promote the euro as an international reserve currency in Asia.

Diana Hochraich takes a more critical view of the large-scale injections of FDI in the Asian countries from the mid-1980s onwards. In Chapter 6, she acknowledges that this FDI did generally aid progress but also argues that it was accompanied by a rapid growth of credit and the money supply, which resulted in inflation, particularly in the stock and property markets, and deterioration in the current accounts of the recipient countries. She suggests that the underlying cause of the crisis was the industrialization strategies followed by the Asian countries, and highlights over-specialization of production, loss of competitiveness, insufficient technology transfer, and an over-reliance on external markets as the crucial flaws in the Asian model. She too draws attention to the limited role played by the European Union in the various rescue packages, and ends with a critical assessment of IMF efforts to promote reform of the affected economies.

The third section presents a number of microeconomic studies, focusing on particular aspects of the relationship between Europe and ASEAN. In Chapter 7, Jim Newton looks at the tourism industry in ASEAN and, in particular, in Thailand. Tourism in Thailand rose markedly in the late 1980s, partly in response to the 'Visit Thailand Year 1987' marketing campaign. The industry now plays a significant role in Thailand's economy, and Thailand has a large share of the European market. Newton therefore queries whether a similar promotional campaign in the late 1990s might also raise tourism, and provide some respite for the crisis-hit ASEAN economies. Unfortunately, he concludes that the tourism industry is unlikely to be able to provide the export-led recovery that its supporters claim, but that industry strategists should plan to diversify to non-traditional markets.

As noted above, the Malaysian Government was among the first of the ASEAN countries to actively promote outward direct investment by indigenous companies through a variety of incentives. The objectives of these policies were twofold: to broaden the earning capacity of the country, and to reduce the large factor payments abroad in the country's balance-of-payments. In Chapter 8, Jim Slater and Isabel Tirado Angel first detail the incentives provided in Malaysia, and then examine the link between ODI and the performance of the firms involved. Their empirical analysis suggests that the incentives offered have had positive effects both on the amount of ODI and on the performance of the investing companies.

Vietnam is one of the newer members of ASEAN, a former centrally-planned economy, and a late convert to economic reform and to the benefits of foreign investment. In Chapter 9, Bernadette

Andréosso-O'Callaghan and John Joyce first outline the process of economic reform and detail the various measures introduced to promote FDI. They then address the question of how transnational corporations (TNCs) choose provincial locations for their affiliates within the country. They demonstrate that TNCs tend to favour the richer provinces, notwithstanding the fact that labour costs are also higher, and also those provinces with a well-educated workforce. In contrast, they find that investment incentives and transport infrastructure have little effect upon the TNCs' location decisions. Thus they conclude that the uneven regional distribution of FDI in Vietnam will endure, despite government efforts to promote a wider spread through the creation of export processing zones and industrial zones.

The fourth and final section looks to the future, and to possible changes in the memberships of the European Union and ASEAN, and in the relationship between the two blocs. In Chapter 10, Richard Pomfret contrasts the possible enlargement of the European Union to include ten countries from Central and Eastern Europe, with that of ASEAN to include Vietnam, Laos, Myanmar and Cambodia. He concludes that ASEAN enlargement has proceeded smoothly because intra-ASEAN trade is of minor importance. In contrast, he foresees great difficulties in EU enlargement because the potential new members would be competing in highly protected EU industrial sectors. The overall result will be enhanced EU–ASEAN relations, with ASEAN's negotiating strength being increased though the balance of power will still remain with the European Union in the foreseeable future. Finally, in Chapter 11, Jim Slater draws together the main conclusions reached and speculates about the future prospects for EU–ASEAN trade and investment.

NOTES

1. ASEAN was founded in 1967 by Indonesia, Malaysia, the Philippines, Singapore and Thailand. Brunei joined in 1984, Vietnam in July 1995, and Laos and Myanmar in July 1997.
2. Singapore, Indonesia, Malaysia and Thailand. The other four HPAEs were Japan, Hong Kong, the Republic of Korea, and Taiwan (World Bank, 1993).

3. The European Union as such was born on 1 November 1993 when the Maastricht Treaty came into effect. Previously the Union had been known as the European Community. However, the term 'European Union' is used throughout this volume to refer to the European Community/Union.

BIBLIOGRAPHY

Commission of the European Communities (1994) *Towards a New Asia Strategy*, COM(94) 314 final (Luxembourg: Office for Official Publications of the European Communities).

World Bank (1993) *The East Asian Miracle* (New York: Oxford University Press).

Section I
The Macroeconomic Picture

1 Bridging Two Continents: EU–ASEAN Trade and Investment

Robert C. Hine

INTRODUCTION[1]

The Association of Southeast Asian Nations (ASEAN), with its combined population of well over 300 million, has become both a major trade partner for the European Union (EU) and a commercially interesting location for foreign direct investment (FDI). This has been achieved in the face of a relatively unfavourable EU trade policy. Recently, there has been a limited re-appraisal of the EU approach in view of the growing economic importance of the ASEAN countries and more generally of the Asia–Pacific region. Compared with other East Asian countries, Europe has considerable historical, cultural and commercial links with most of ASEAN's member countries. ASEAN could therefore be seen as a bridge for Europe to the wider Asia–Pacific region. The United States (US) has such a link through its participation in the Asia–Pacific Economic Cooperation (APEC) forum. Should the European Union thus seek to strengthen its institutional link with ASEAN and adopt a more open and supportive trade policy framework? A difficulty for EU policymakers has been that exports from ASEAN are highly competitive in European markets, provoking protectionist demands and making it difficult to maintain open market access. For their part, the ASEAN countries are focused on the short-term imperatives of dealing with the financial and economic crisis that struck them in 1997. Hence, the EU–ASEAN bridge instead of being enlarged is in danger of being blocked. As blockage is more likely the more disparate is the traffic in each direction and the more restrictive are the rules, this chapter examines the nature of EU–ASEAN trade and investment flows, their adjustment implications and the EU's trade treatment of ASEAN.

The chapter is structured as follows. The next section sets the context of the rapid development of the ASEAN economies prior to 1997, and especially the role played by foreign trade and investment. In the third section, the ambivalent nature of EU trade policies towards the ASEAN countries is explored. This leads in the following section to an analysis of EU–ASEAN trade patterns emphasizing patterns of specialization and intra-industry trade. The impact of the Asian financial crisis on the bilateral economic relationship is considered in the penultimate section, and a final section concludes.

ASEAN AND THE 'EAST ASIAN MIRACLE'

Four of the founding members of ASEAN – Indonesia, Malaysia, Singapore and Thailand – were included in an elite group of eight 'high-performing Asian economies' (HPAE), which constituted the driving force of the World Bank's 'East Asian Miracle'. The HPAE achieved an annual growth in GNP per capita of 5.5 per cent over the period 1965 to 1990 – more than twice the growth rate in the OECD economies (World Bank, 1993: 2). The recipe for this success is in dispute. It certainly included a combination of: high savings and investment rates, a relatively high degree of equality, high growth rates of human and physical capital, high productivity growth, high growth in manufactured exports, high growth in agricultural productivity and a decline in fertility. According to the World Bank, governments contributed importantly, especially by maintaining macroeconomic discipline. However, in other respects – such as interventionist industrial policies and restrictive import regimes – their impact was at best neutral (Morrissey and Nelson, 1998).

Beginning with Singapore in the 1960s, the ASEAN countries achieved a gradual transformation of their economies from largely agricultural to substantially industrial. The economic philosophy of an inward-looking industrialization based on import substitution gave way to an approach that focused on manufactured exports as the engine of growth. Particularly in the 1970s, there was a sharp increase in the importance of trade in the ASEAN economies – in Thailand and Indonesia the ratio of trade to GDP almost doubled in the decade (World Bank, 1993). Encouragement to exports was, however, generally combined with continued high barriers to imports, except

Table 1.1   Simple Average Applied Tariffs of the ASEAN
Countries, 1984–96

Country	Average Applied Tariff		
	1984–87	*1991–93*	*1996*
Thailand	31.2	37.8	17.0[1]
Philippines	27.9	23.5	15.6
Indonesia	18.1	17.0	13.4
Malaysia	13.6	12.8	9.0
Brunei	3.9	3.9	2.0
Singapore	0.3	0.4	0.0[2]

Notes: (1) 1997 tariff.
(2) Excludes bound specific duties on agricultural
products.
Source: Choy (1998).

in Singapore where 96 per cent of products enter duty-free. Even
with recent tariff cuts negotiated in the Uruguay Round, the average
level of nominal tariffs in the ASEAN countries remains high – see
Table 1.1.

The growing exposure of the ASEAN economies to trade was
accompanied by a stable share of ASEAN (excluding Singapore) in
world trade from 1961 to 1976. But as the region's economic growth
accelerated, its share of world trade almost doubled, rising from 3.3
per cent in 1980 to 6.4 per cent in 1995.[2] The five founder members
of ASEAN were ranked among the world's fastest growing exporters
in 1990–96. While world trade grew at an average annual rate of
7 per cent during this period, exports of the ASEAN countries rose
by 12–18 per cent. This growth spurt has left four of the ASEAN
countries in the top thirty of international traders: Singapore (the
13th largest exporter in 1997), Malaysia (18th), Thailand (22nd) and
Indonesia (25th) (WTO, 1998: 6). The economic transformation of
the ASEAN economies is mirrored in the commodity pattern of their
exports. Primary products have been replaced by manufactures as
the dominant export category – manufactures account for some 80
per cent of export earnings. However, the ASEAN countries continue to play a leading role in some primary product markets. These
include tropical timber, rubber, tin and other metals and energy.

ASEAN countries, particularly Singapore, are also important traders in services. ASEAN's main trade links outside Asia are with the United States and the European Union. They are of similar importance as sources of imports (15 per cent each), with the United States a little ahead as an export destination (19 per cent compared with 14 per cent for the European Union). But the majority of ASEAN merchandise trade is with other Asian countries. The largest economy in the region, Japan, accounts for 15 per cent of exports and 19 per cent of imports (IMF, 1996). Other developing countries in Asia, including partners in ASEAN, account for a further 42 per cent of exports and 33 per cent of imports. Intra-ASEAN trade has been encouraged since 1977 by a Preferential Trade Agreement but, at least initially, to minimal effect (Panagariya, 1994). This is unsurprising since, according to Pomfret (1993), by 1981 only 2 per cent of intra-ASEAN trade was covered by preferential tariff reductions, and even by the end of the 1980s the share was only 5 per cent. Member countries were reluctant to expose their infant industries to external competition, and there was limited scope for trade among a group of developing countries with similar export interests. Over the last decade, intra-ASEAN trade appears to have grown rapidly, more than matching the growth of trade with non-member countries.[3] However, much intra-ASEAN trade comprises non-ASEAN products, especially crude oil and petroleum, re-exported to ASEAN through Singapore as entrepôt. Indeed, if crude oil and petroleum products are omitted, intra-ASEAN trade is only 5 per cent of total ASEAN exports. Evidence from Diwan and Hoekman (1999) confirms the limited role of intra-ASEAN trade (except for the strong link between Malaysia and Singapore), but shows that bilateral export intensities are nevertheless two to three times higher than would be expected based on world trade shares.

ASEAN's export surge in the 1980s and 1990s was underpinned by a substantial influx of FDI. Malaysia, Indonesia and Thailand, for example, had net private inflows of capital rising from $9.5 billion in 1990 to $47.2 billion in 1996. ASEAN's share of world FDI inflows before the 1997 crisis was about 6 per cent[4] – well ahead of its share in world GDP (3.5 per cent), but in line with its share of world trade (WTO, 1998: 5).[5] Traditionally, the European countries were the dominant source of foreign investment in ASEAN. However, since the 1970s Japan has taken the leading role. In the 1980s in particular there was a wave of inward investment from Japan and other east Asian countries. By 1993, Japan's share of FDI stocks in ASEAN (excluding Brunei and Singapore) was 22 per cent, compared with 15 per cent for the European Union and 10 per cent for the United States. The shares for Singapore were, respectively, 32 per cent, 27 per cent and 35 per cent. Much Japanese investment has been driven by production cost considerations – especially following the appreciation of the yen in the second half of the 1980s. FDI has led to intensified intra-industry trade between Japan and the ASEAN countries, both in components and outputs. The interlinking of the East Asian economies through trade and investment played an important role in spreading economic growth through the region, but also increased the vulnerability of countries to a downturn in neighbouring economies.

For EU investors, Singapore has historically been the most important host country in ASEAN, but Indonesia, Malaysia and Thailand also attract substantial European FDI. The activities receiving the largest flows are petroleum extraction and refining, chemicals and pharmaceuticals, processed foods, electrical and electronic products, automobile assembly, and banking and finance (Commission of the European Communitities, 1997). Besides the tariff-jumping motivation for firms wishing to sell in the ASEAN market, ASEAN is seen as a low cost manufacturing base by European, as well as Japanese, producers. This includes higher technology products, particularly electronics, for export to the US and third country markets.[6] In the reverse direction, ASEAN flows of FDI to the European Union are small but increasing – see Chapter 4 for further details. Before the Asian crisis, the European Union attracted about 10 per cent of ASEAN outward investment, much of it in services. The main motivation for this FDI is probably strategic – to gain access to European technology and know-how.

In short, the ASEAN economies until 1997 experienced a period of spectacular growth based on rapid industrialization. A high proportion of manufactures was for export, and industrial growth was underpinned by a network of FDI both regional and global. It is important to emphasize, however, that the ASEAN economies still remain very diverse. Singapore is the most advanced, but at the other end of the spectrum, the new members – Vietnam, Laos and

Myanmar – are at a very early stage in the industrialization process. In the 'flying geese' model (Akamatsu, 1962), countries pass through successive patterns of specialization as their economies develop. This brings them into competition with different groups of countries. Hence Vietnam, for example, faces intense low wage competition from China in clothing and textiles, whereas competition for the medium-income ASEAN countries comes more from Latin America and central and eastern Europe, and the sophistication of the Singaporean economy places it on a comparable level to Japan.

EU TRADE POLICIES TOWARDS ASEAN

Although the EU countries have strong historical ties with most of the ASEAN countries, institutional links between the two regional groupings have been weak. A Cooperation Agreement between the European Community (EC) and ASEAN was signed in 1980 with the aim of providing a forum for political and economic dialogue, but no major progress was made until the early 1990s. The EC's preoccupation with internal developments and with its trade partners elsewhere, the relatively low purchasing power of the region until recently, and the lack of progress in economic integration within ASEAN all played a part in this neglect. Changing circumstances have prompted a limited re-appraisal of relations in the European Union. In particular, the region has become an increasingly important destination for EU exports (Hill and Phillips, 1993) – ASEAN's share of EC exports rose from 3 per cent in 1980 to 6.5 per cent in 1996. Also, the growing prominence of the Asian region in the world economy and the activities of APEC – whose economies account for a half of world output and trade – has elicited an increased interest in ASEAN as a gateway to the Asia–Pacific region. In its 'New Asia Strategy' (Commission of the European Communities, 1994), the European Union recognized the benefits of closer interaction with the dynamic Asian economies and the need to integrate the former State-controlled economies in the region more fully into the world economic system. The most concrete evidence of the new approach is the annual Asia–Europe Meeting (ASEM) inaugurated in 1996 and attended by representatives of the European Union and Asian countries including ASEAN, Japan, China and Korea. The meetings provide a basis for strengthening economic cooperation between the participants. They include work on practical if unspectacular areas such as patents and trademarks.[7] The 1980 Cooperation Agreement and the launch of ASEM cannot however disguise the fact that ASEAN has a low ranking in the EU's system of external relations – it is one of the EU's least privileged trade partners. Moreover, the EU's trade policies towards ASEAN, as with other groups of developing countries, are ambivalent. On the one hand, the ASEAN countries benefit from preferential access to the EU market under the Generalized System of Preferences (GSP). On the other hand, regimes like the Multi-Fibre Arrangement (MFA) have held back competitive exports from the ASEAN countries, and anti-dumping duties have been extensively used against imports from Asia. Thus the European Union's willingness to support the economic development of low and middle-income countries through liberal access to the huge EU market is qualified by an overriding concern not to aggravate the adjustment pressures of declining EU industries.

Tariffs are no longer a major obstacle to access to the EU market for most products. Following the implementation of cuts agreed in the Uruguay Round, the weighted average tariff on manufactures will be around 4 per cent in 2000. Tariffs on a range of products have been eliminated altogether but some relatively high tariffs remain, particularly on certain products of export interest to the ASEAN countries such as microprocessors, radio and television sets, and clothing and textiles. This is of course not accidental – in multilateral negotiations the European Union has resisted cutting tariffs on imports where EU industry is facing a strong challenge from imports from the Newly Industrializing Countries. ASEAN suppliers also face discrimination in that the EU offers more favourable treatment to rival producers in other regions – in Central Europe (Europe Agreements), Africa (Lomé Convention), and the Mediterranean (Euro–Med Agreements).

This discriminatory treatment is only partially reversed through the GSP. Most manufactured products (excluding those covered by the MFA) obtain tariff concessions under the GSP, but their extent depends on the 'sensitivity' of the product. Sensitivity reflects both the adjustment pressures in the EU import-competing industry, and its political influence in this autonomous – i.e. non-negotiated –

arrangement. For *sensitive* products, the MFN tariff is cut by just 20 per cent; this increases to 60 per cent for *semi-sensitive* products, while for *non-sensitive* products there is tariff-free access. In addition, there is a formula for denying tariff preferences for a particular product from a country that has exceeded a threshold share of the EU market, depending also on the national per capita income. The stated aim of this arrangement is to encourage new producers to develop exports of the product. This would also be the result of graduating certain countries that have reached a threshold income level. Thus, among the ASEAN countries, Singapore no longer qualifies for GSP preferences. A further deterrent to GSP usage is compliance with 'rules of origin', though for ASEAN there is at least the concession that origin is judged in relation to the area as a whole and not just the territory of the exporting country. In practice, only a minority of 'GSP eligible' exports actually receive preferential treatment in the European Union.

Textiles and clothing exports from ASEAN to the European Union have been subject to trade restraints under the MFA since 1974. In addition to the already high tariff rates on these products, access to the EU market is limited by quotas, set for each of a large number of product groups and for each ASEAN country that is a significant supplier of the product. Although the quotas have been progressively expanded, the MFA has held back the growth of developing country exports. It has also helped to fossilize the pattern of trade, sometimes benefiting the more advanced developing countries whose quota allocations reflected their former comparative advantage in the sector. Under the terms of the Uruguay Round agreement, the MFA is being phased out with more products being exempted from quota and quota amounts increased. The regime is planned to end in 2005.

The EU has deployed other instruments to blunt the competitive challenge from the ASEAN countries. 'Voluntary' export restraints (VERs) have been applied on Singapore's exports of colour televisions to the United Kingdom, and manioc exports from Thailand; but such actions are now prohibited within the World Trade Organization (WTO). Critics consider that anti-dumping (AD) measures are now the primary form of new protection in the European Union (Messerlin and Reed, 1995). Under WTO rules, products are considered dumped if (a) they are sold in the export market at a price which is below their normal value in the exporting country, and

(b) this causes, or threatens to cause, material injury to producers in the importing country. There is evidence that the AD regime in the European Union has been captured by producer organizations that are able to influence the outcome of the AD investigation. The significance of this is that very high rates of AD duty have been imposed on selected suppliers. Moreover the duties remain in force for five years, with a possible five-year extension. ASEAN countries have been particular targets of EU anti-dumping actions. For example, between 1992 and 1996, Thailand was the second most frequently affected country in terms of the number of AD investigations launched by the European Union, while Indonesia and Malaysia held joint fourth place. Since 1995, the number of products on which AD duties have been definitively imposed has increased sharply and there was a very large upsurge in 1998 in the wake of the Asian crisis – see Table 1.2. The duties imposed have ranged up to 61 per cent.

Thus although the European Union has made some concessions on ASEAN exports under the GSP, these are insecure since ASEAN export success is likely to result in the loss of preference. Moreover EU tariffs tend to be relatively high on ASEAN products, and reinforced in the case of clothing and textiles by quotas. An increasing number of products are affected by anti-dumping duties. This relatively unfavourable trade policy environment reflects in part the

Table 1.2 The Numbers of EU Anti-Dumping Cases involving the Imposition of Definitive Duties on Imports from the ASEAN Countries, 1990–98

Year	Items (8 digit tariff line)	Related Product Groups
1990	2	2
1991	2	2
1992	6	1
1993	1	1
1994	0	0
1995	18	3
1996	14	9
1997	14	4
1998	42	12

Source: *Official Journal of the European Union* (various issues).

nature of the trade relationship which is examined in the next section. However, it should also be pointed out that ASEAN import arrangements, with the notable exception of Singapore, are also very restrictive for many products.

AN ANALYSIS OF EU–ASEAN TRADE

EU–ASEAN trade quadrupled in value over the last decade. This has helped to reduce the asymmetry in the economic relations of the two blocs. Nevertheless, there is still an important imbalance – the European Union accounts for about 15 per cent of the ASEAN countries' trade, whereas the ASEAN market accounts for only about 6.5 per cent of EU external trade. There is also a divergence in commodity structures since the ASEAN countries have traditionally been important exporters of tropical agricultural products and raw materials. But the value of these exports to the European Union has been stable over the last decade at about £2.5 billion. Thus agriculture and mining's share of EU imports from ASEAN has fallen from 20 per cent in 1988 to 8 per cent in 1996. This reflects the declining share of primary production in the ASEAN economies, especially in Thailand and Malaysia where agriculture's contribution to GDP approximately halved during the growth spurt of the late 1980s and early 1990s.

The great bulk of trade between ASEAN and the European Union in both directions is in manufactured goods. The main areas of EU exports to ASEAN are electronic and electrical equipment (including telecommunications), machinery for industry, and transport equipment – motor vehicles and aircraft. Seven industries (out of 134)[8] account for 40 per cent of EU exports to ASEAN. In the reverse direction, ASEAN exports to the European Union are even more concentrated by industry – just two accounted for 40 per cent of ASEAN's exports to the European Union in 1996 – office and data processing machines (21 per cent) and electronic equipment, radios & televisions (19 per cent). Schmitt-Rink and Lilienbecker (1991) noted the build-up of computer exports from ASEAN to the European Union during the second half of the 1980s, as European companies transferred the production of electronic components and their assembly to Southeast Asia. Today, Singapore plays the leading role in exports of office and data processing equipment and tele-communications. Malaysia has a particularly strong presence in exports of electrical equipment – accounting for about half of the ASEAN total. The Philippines also has a small but growing electronics sector. More traditional labour-intensive industries such as hosiery, clothes and footwear currently play a relatively small role in ASEAN exports – each accounts for only about 3 per cent of ASEAN exports to the European Union. Thailand supplied about half of ASEAN's footwear exports to the European Union in 1988, and 40 per cent of clothing. But Indonesia's share of the footwear, clothing and textiles exports has grown substantially over the last decade – by 1996 it accounted for 60 per cent of footwear and 50 per cent of clothing and textiles exports from ASEAN.

Differences in trade specialization between ASEAN and the European Union are illustrated by the index of revealed comparative advantage (RCA):

$$RCA = (x_i/X) - (m_i/M)$$

where x_i and m_i are, respectively, EU exports to ASEAN and imports from ASEAN of products in industry i; and X and M are, respectively, total EU exports to ASEAN and total EU imports from ASEAN.

At a fairly aggregated level (28 industries), the RCA analysis indicates an EU strength in the machinery, chemicals and transport equipment sectors and a weakness in natural-resource based industries and simple manufactures such as textiles and clothing and footwear – see Table 1.3. Over the last decade there has been a general convergence in the bilateral trade pattern so that RCA is becoming less distinct – thus, for example, the EU's positive index for chemicals has halved over this period. The exceptions to this trend towards convergence are at the two extremes. The EU's positive RCA in agricultural and industrial machinery has strengthened, whilst the negative RCA for office and data processing equipment has intensified very strongly. This pattern is confirmed when the RCA analysis is carried out at a more disaggregated level (138 industries): the predominant tendency is towards a convergence of the industry structure of imports and exports.[9] Again the main exceptions were, for positive RCAs, machinery sectors (together with transport equipment and some electrical goods) and, for negative

Table 1.3   EU Revealed Comparative Advantage in Trade with the
ASEAN Countries, 1988 & 1996

NACE– CLIO	Industry	RCA 1988	RCA 1996
21	Agricultural and industrial machinery	0.176	0.217
17	Chemical products	0.134	0.061
27	Motor vehicles	0.045	0.051
29	Other transport equipment	0.043	0.040
13	Ferrous & non-ferrous ores & metals	0.039	0.021
19	Metal products ex machinery & transport equipment	0.026	0.019
15	Non-metallic mineral products	0.020	0.016
47	Paper & printing products	0.015	0.012
37	Beverages	0.020	0.012
33	Milk & dairy products	0.014	0.006
55	Scrap metals, waste paper etc.	0.001	0.005
39	Tobacco products	0.003	0.004
31	Meats & meat preparations	0.001	0.000
51	Other manufacturing products	0.024	0.000
07	Crude petroleum, natural gas & petroleum products	0.002	−0.002
03	Coal	0.000	−0.003
25	Electrical goods	0.011	−0.010
49	Rubber & plastic products	−0.003	−0.014
43	Leather, leather & skin goods, footwear	−0.016	−0.026
45	Timber, wooden products & furniture	−0.108	−0.043
35	Other food products	−0.097	−0.054
01	Agricultural, forestry & fishery products	−0.192	−0.054
41	Textiles & clothing	−0.117	−0.074
23	Office & data processing machines; precision & optical instruments	−0.050	−0.183

Notes: (1) RCA = revealed comparative advantage.
(2) NACE = Nomenclature Genérale des Activités Economiques
dans les Communautés Européenes; CLIO = Conversational
Language for Input–Output.
(3) The following NACE–CLIO 2-digit categories had RCA of
0.000 in both years: 05 'Coking products'; 09 'Electric power,
gas, steam and water'; 11 'Radioactive materials & ores'; 79
'Market recreational & cultural services'.

Source: Choy (1998).

RCAs, office and data processing equipment (plus some simple manufactures such as footwear, games and toys). The electrical/ electronics sectors thus appear to be unusual in experiencing growing specialization – the ASEAN countries in data processing and office equipment, radios and televisions; the European Union in telecommunications and electric motors.

Standard trade theory suggests that patterns of specialization reflect relative resource endowments. Countries specialize in, and export, goods whose production uses intensively their abundant factors of production. On this basis, the less developed ASEAN countries should specialize in unskilled labour-intensive industries[10] such as footwear, while the EU's strength should be in capital-intensive areas, particularly those that require a high input of skilled labour. A simple test of this in relation to EU–ASEAN trade is to compare the pattern of RCA across industries with the industry pattern of factor usage. A measure of the latter is the value-added at factor cost per person employed:[11] higher value-added should reflect a greater capital–labour ratio and hence capital intensity. For 92 manufacturing industries (again defined at the three digit NACE level), the Spearman rank-order correlation coefficient between the RCA and value-added per employee is consistently above +0.4 for the years between 1988 and 1996,[12] reaching a peak of +0.5 in 1993. This is consistent with the view that the European Union's greater relative endowment of capital, particularly human capital, gives it a comparative advantage in the more capital-intensive sectors of manufacturing industry. High rates of investment in ASEAN would erode this advantage over time, and indeed until the crisis there was evidence of a general convergence in EU–ASEAN trade patterns.

This convergence also implies that inter-industry (net) trade between the two blocs will become less important than intra-industry trade (IIT). That is, trade will shift from being one-way in a particular industry towards a simultaneous export and import of products of the same industry. The main drivers for the latter type of trade are generally seen as product differentiation in response to consumer demands for variety, and economies of scale in production. The growth of IIT has implications for adjustment and trade policy since, in principle, IIT could expand rapidly without necessitating an extensive redeployment of labour. With the same labour force, a higher proportion of production could be exported, whilst domestic consumers buy an increasing proportion of foreign products. An expansion of IIT could therefore cause fewer trade frictions than if labour has to be switched between declining import-competing industries and expanding export industries. It is therefore useful to examine the extent to which EU–ASEAN trade is based on IIT.

The standard measure of the share of IIT in total trade is the Grubel–Lloyd (GL) index:

$$B_j = [1 - (X_j - M_j)/(X_j + M_j)] \times 100 \quad (0 \le B_j \le 100)$$

where X_j are exports of industry j and M_j are imports of industry j.

The extent of IIT across a range of industries is typically reported using a weighted average of the IIT of each industry, using the share of each industry in total trade as the industry weights.[13] Table 1.4 reports the Grubel–Lloyd index for EU–ASEAN trade at the four-digit level of the Combined Nomenclature. The main finding is that only a minority of EU–ASEAN trade is IIT – for the most part the European Union and ASEAN export different products to each other. The GL index has been relatively stable over the last decade.

However, there are large differences between the ASEAN countries – Indonesia and Vietnam, in particular, have very low levels of IIT – see Table 1.5. The generally low level of IIT suggests that the specialization gains from trade expansion may be accompanied by adjustment pressures in the European Union. A possible mitigating factor is that ASEAN RCA increases were mainly in sectors characterized by rapid demand growth. Hence the impact in these sectors may have been to hold back job creation in the European Union, rather than to displace existing workers.

There is a sense in which even the low observed level of IIT may understate the adjustment pressures experienced by the European Union. The growth of IIT will only have a benign effect on adjust-

Table 1.4   The Grubel–Lloyd Index for EU–ASEAN Intra-Industry Trade, 1988–95

Year	1988	1989	1990	1991	1992	1993	1994	1995
GL index	25.1	26.3	27.0	25.9	24.4	25.2	27.2	28.4

Source: Author's calculations based on EUROSTAT 4-digit data.

Table 1.5   The Composition of EU–ASEAN Intra-Industry Trade, 1995

Country	Total Intra-Industry Trade (ecu billion)	Grubel–Lloyd Index			
		Total IIT	HIIT	VHIIT	VLIIT
Singapore	11128	36.2	2.5	17.8	15.9
Malaysia	8187	32.5	1.8	22.5	8.1.
Thailand	4331	23.5	1.5	14.1	7.8
Philippines	1719	33.3	0.7	28.1	4.5
Indonesia	1267	9.0	0.8	6.7	1.5
Brunei	368	62.0	54.2	1.1	6.7
Vietnam	65	6.4	0.5	4.8	1.1
Total (1995)	27065	28.4	2.1	17.0	9.3
Total (1988)	6742	25.1	6.4	12.6	6.1

Notes: IIT = Intra-Industry Trade; HIIT = horizontally-differentiated IIT; VHIIT = vertically-differentiated IIT, where the unit value of exports is 15% or more higher than the unit value of imports; VLIIT = vertically-differentiated IIT, where the unit value of exports is 15% or more lower than the unit value of imports.
Source: Author's calculations based on EUROSTAT 4-digit CN data.

ment if the products being exchanged are of similar qualities and hence are produced using similar factor combinations. Otherwise, shifting resources out of low quality into high quality (or vice versa) branches of the same industry may result in a net loss of some factors and a net influx of others. Schmitt-Rink and Lilienbecker (1991) found evidence that, in the 1980s, EU–ASEAN trade in electronics was vertically differentiated, both through vertical product differentiation (audio/video equipment) and through vertical process specialization (technical electronics such as power equipment). Hence it is useful to examine further the nature of IIT between ASEAN and the European Union to establish the extent to which IIT is vertically differentiated.

A method of distinguishing vertical from horizontal differentiation in IIT is to compare the unit values[14] of imports and exports, on the assumption that higher unit values will in general reflect higher qualities (see Greenaway *et al.*, 1994, 1995). Hence, where the unit values of imports and exports in a disaggregated product category differ by more than a certain value (say 15 per cent, to allow for transport cost effects etc.) then the IIT could be regarded as verti-cally-differentiated (VIIT) – the exchange of different qualities of the same product or group of products. Where unit value differences are below the threshold value, IIT can be regarded as horizontally-differentiated (HIIT). VIIT can be further subdivided into IIT where EU exports are of higher unit value than imports (VHIIT), and IIT where EU import unit values exceed the corresponding export unit values (VLIIT).

Table 1.5 shows that IIT is overwhelmingly vertical for manu-facturing, which accounts for virtually all of the IIT in EU–ASEAN trade. This is consistent with the findings of empirical studies for other countries (e.g. Greenaway *et al.*, 1995). But the proportion of HIIT is particularly low in EU–ASEAN trade – about 15 per cent. Given that quality differences in VIIT are driven by differences in factor endowments, high quality varieties should be exported by countries having a relative abundance of capital. Hence, it would be expected that for the EU trade with the ASEAN countries, VHIIT would predominate over VLIIT. This is indeed confirmed by the results shown in Table 1.5. Even where there is two-way trade between the European Union and ASEAN in products of the same industry, the evidence summarized in the table suggests that there are important quality differences. Hence, an expansion of IIT could still generate significant adjustment pressures.

The analysis thus far has focused on *relative import–export* unit values. From a trade policy perspective, it is also useful to compare the unit values in trade between the European Union and ASEAN with unit values within the European market. How does the quality of imports from ASEAN compare with those traded inside the EU? Are significant changes in this relationship occurring as the ASEAN economies develop? As with the earlier analysis, threshold values of ±15 per cent around the intra-EU average can be used to categorize imports and exports from ASEAN. However, the analysis is now extended from just intra-industry trade to all trade between the European Union and ASEAN. On this basis, there is a clear tend-ency for ASEAN exports to the European Union to have low unit values – about 50 per cent of imports have unit values below 0.85 of the intra-EU value, and 20 per cent are within the 0.85 to 1.15 band. Insofar as unit values reflect quality, imports of ASEAN products appear to be of lower quality compared with European products classified into the same 4-digit trade category. Alternatively, the low unit values could point to ASEAN products being sold at a discount in order perhaps to establish the product in the EU market place and to build up market share. From a trade law perspective, this would constitute dumping only if the price in ASEAN were higher than the export price. Another possibility is that the low unit values reflect transfer pricing practices within multinational companies. Fluctuating exchange rates might also contribute to differences in unit values at a given moment, but the differences noted have been very persistent.

In the reverse direction, the unit values of EU exports to ASEAN also display an interesting characteristic. Compared with the benchmark intra-EU trade, these exports are biased towards high unit value products – about a half have export unit values that are 15 per cent or more above the intra-EU benchmark. What might account for this bias? One possible explanation is that trade barriers in ASEAN have a greater deterrent effect on the lower-quality EU exports than the more sophisticated ones – i.e. the price elasticity of demand declines with increasing quality. Hence trade barriers are less of an obstacle for high-quality products. Another possibility is that ASEAN producers have concentrated on simpler manufactures and do not yet have the capacity to produce more sophisticated products such as large passenger aircraft. In the presence of tariffs and transport costs, this would tend to bias EU exports towards more sophisticated products. Whatever the explanation, the pattern has been persistent over the last decade in the face of a large expansion of EU–ASEAN trade.

Thus far, import and export unit values have been compared (a) to each other to establish vertical and horizontal IIT, and (b) relative to the intra-EU benchmark. The final stage is to examine IIT using both measures in order to get a better understanding of what drives vertical IIT. Table 1.6 summarizes the results first using ASEAN import values relative to the intra-EU benchmarks. This shows that the dominant strand of IIT between the European Union and ASEAN is the exchange of high-quality EU exports for low-quality imports – VIIT of high-quality from an EU perspective. However, there is also a substantial amount of IIT where the ASEAN exports are of higher quality, relative to EU exports and relative to the intra-EU benchmark. Second, comparing EU export unit values with the

Table 1.6   The Value of EU–ASEAN Vertical-Differentiated and Horizontally-Differentiated Intra-Industry Trade, 1995

(A) Unit Values of EU Imports from ASEAN relative to intra-EU Unit Values

Ratio	Nature of Intra-Industry Trade			
	Total IIT	*HIIT*	*VHIIT*	*VLIIT*
High (> 1.15)	23	3	5	15
Medium (0.85–1.15)	11	2	8	1
Low (< 0.85)	25	2	23	0
Total	59	7	36	16

(B) Unit Values of EU Exports to ASEAN relative to intra-EU Unit Values

Ratio	Nature of Intra-Industry Trade			
	Total IIT	*HIIT*	*VHIIT*	*VLIIT*
High (> 1.15)	37	3	28	6
Medium (0.85–1.15)	14	2	6	6
Low (< 0.85)	10	2	3	5
Total	61	7	37	17

Notes: IIT = Intra-Industry Trade;   HIIT = horizontally-differentiated IIT; VHIIT = vertically-differentiated IIT, where the unit value of exports is 15% or more higher than the unit value of imports; VLIIT = vertically-differentiated IIT, where the unit value of exports is 15% or more lower than the unit value of imports.
Source: Author's calculations based on EUROSTAT 4-digit CN data.

corresponding intra-EU values, EU exports are shown to be very largely of medium or high-quality relative to the EU benchmark. The general absence of low-quality EU exports to ASEAN results in an asymmetry in vertical IIT.[15]

This section has highlighted some important differences in bilateral EU–ASEAN trade. Although the great bulk of intra-bloc trade is now in manufactures, the products traded in each direction show considerable differences – the European Union for example has a strong export position in specialized machinery whereas ASEAN's specialization is in electronic and electrical goods. Unlike the situation within the European Union, there is relatively little intra-industry trade, and even where two-way trade exists there are typically marked differences in unit values. This suggests that different qualities of product are being exchanged. The policy implications are twofold: on the one hand, EU–ASEAN trade generates important gains from increased specialization and competition; on the other hand, with a rapid expansion of trade, adjustment problems could lead to protectionist pressures in the European Union.

IMPLICATIONS OF THE ASIAN FINANCIAL CRISIS

In the second half of 1997 a major crisis engulfed the ASEAN economies. It was triggered by a number of factors: the beginnings of cyclical weakening, an unexpected exchange rate policy change, and monetary tightening. There followed a sharp fall in asset prices, generating acute financial difficulties and with contagion between markets. Stock markets slumped, reducing net worth and borrowing capacity. Borrowers were hit further by intensified credit rationing, and there were runs on banks. The loss of liquidity in securities markets served to lock investors in, reinforced in Malaysia by the imposition of exchange controls. The extent of the crisis reflected underlying problems in the ASEAN economies. Their previous fast growth had been based on high rates of investment which initially was highly profitable. Later on, returns diminished, partly due to 'catch-up' but also due to misallocation of resources (e.g. to prestige projects and excessive real estate development). The spurt of development in the 1990s led to a rapid build-up of debt and rising leverage. There was inadequate credit assessment not only by local banks but also international portfolio investors and banks. Foreign investors may have been lulled into believing that they could at any time withdraw funds without loss while some foreign banks may have been overaggressive in seeking expanded market shares. Thus 'the actors in the situation developed a form of disaster myopia, with warning signs such as current account deficits, loss of real competitiveness and evident misallocation of capital being disregarded' (Davis, 1998).

Diwan and Hoekman (1999) argue that the contagion may have been amplified by trade and investment patterns. The 'competition-cum-export-similarity' story suggests that the ASEAN economies have faced a competitive squeeze in recent years. At the bottom end of the market, this comes from the emergence of China as a major exporter of unskilled-labour-intensive products, forcing the poorer ASEAN countries to move prematurely up the technology ladder and leading to a strong demand for investment including FDI. At the higher end of the market, the depreciation of the yen increased competition from Japan for the more sophisticated manufactures. Many East Asian countries then tried to scramble back towards more medium-technology products. Thus rivalry amongst the Asian economies spread the crisis through the region.

The alternative story considered by Diwan and Hoekman is that the magnitude and speed of the contagion reflected the extent of regional integration – the Asian economies were complementary and interdependent so that a major problem in one country had an impact throughout the region. The critical player here was Japan because of its position as an important source of technology, financial capital, capital goods and a large market for East Asian output. Analysis of trade and investment patterns led Diwan and Hoekman to conclude that the complementarity of trade and investment was a more important factor in contagion than the competition-cum-export-similarity argument. The positive aspect of this is that over time the depreciation of the Asian currencies will enhance the competitiveness of the Asian 'joint product' on world markets. Indeed, Asian exports grew strongly in volume in 1997 and 1998, while imports into the region declined. Sustaining this positive story does however depend on a recovery in the Japanese economy. Meantime, FDI flows to the region have plummeted. Net private flows to Indonesia, Korea, Malaysia, the Philippines and Thailand having leapt from $48 billion in 1994 to $93 billion in 1996, then collapsed to minus $12 billion in 1997.[16]

From an EU perspective, the economic impact of the Asian crisis has been limited. A previously very buoyant export market has – albeit temporarily – been set back, and investors have had their fingers burnt. But action by the IMF helped to prevent a regional crisis from becoming a global one. Moreover, the Asian storm cloud had a silver lining for the European Union and the United States in that it reduced inflationary pressures in the world economy. However, the devaluation of the Asian currencies increased the competitiveness of imports into Europe generating stronger protectionist pressures.

ASEAN's response to the crisis has been to push ahead with plans for integration. In 1992, it had been agreed to establish an ASEAN free trade area (AFTA) in manufactured goods over a 15-year period. Although the impact on trade was expected to be limited (Imada, 1993; Ariff, 1995), it might have important knock-on effects in reducing MFN tariffs, helping to reduce trade diversion effects, and encouraging further FDI (Authukorala and Menon, 1997). By 2000, it is expected that 85 per cent of tariffs on intra-area trade will be 5 per cent or less.[17] In a bid to prevent the crisis from damaging ASEAN trade, it was agreed in December 1998 to implement AFTA from 2002, a year ahead of the revised schedule. ASEAN countries have also agreed to offer new tax incentives to encourage a revival of investment.

CONCLUDING REMARKS

International trade and investment continue to play a vital role in the development of both the EU and the ASEAN economies. They have contributed to a high level of specialization and generate a beneficial competitive stimulus. But managing globalization can be problematic, as vividly exemplified by the crisis which hit the Asian region in 1997. Less acutely, trade has generated adjustment problems for the European Union, particularly in the declining industries some of which are also geographically concentrated within Europe. In this process, the nature of trade patterns is crucial. Within the European Union, trade-induced adjustment pressures appear to have been subdued because of the complex pattern of two-way trade in similar products. Different products tend to be traded in each direction between the European Union and ASEAN, reflecting different patterns of comparative advantage. Moreover, there are important differences in unit values between EU imports and exports, suggesting that ASEAN countries are specializing in different segments of the market. Hence trade expansion with ASEAN may generate significant adjustment pressures, leading to interest group pressures for protection. This is reflected in trade arrangements – for example, 'sensitive' product treatment within the GSP and, arguably, the use of AD measures.

The rapid expansion of the ASEAN economies before 1997 led to a limited re-appraisal of EU policy. The European Union was anxious to benefit from the dynamic market in ASEAN for its exports. Moreover the establishment of APEC aroused concerns that the European Union would face discrimination in a major portion of the world market, particularly in relation to the United States. Thus the European Union has attempted to build bridges with the Asian economies, particularly through ASEAN. ASEM is a concrete part of this process. The danger is that short-term adjustment issues could dominate longer-term strategic issues preventing the EU–ASEAN bridge from developing effectively. The Asian crisis adds a new twist to this debate. On the one hand, the economic slowdown in Asia could cause the European Union to lose interest in developing stronger Asian links. Attention could re-focus on Europe's more immediate neighbours, especially following the Balkan conflict. On the other hand, both the European Union and the United States have an important long-term interest in the recovery of the Asian economies. The Asian market is crucial for many exporters, and many EU firms now also have a substantial investment stake in the region. Regional recovery is likely to involve a further expansion of exports to the European Union and the United States. In this scenario, the EU–ASEAN bridge has an important role to play.

NOTES

1. The research assistance of Ian Gillson and Isabel Choy is gratefully acknowledged.
2. These shares are inclusive of Singapore.
3. The share of total ASEAN exports sold within the grouping rose from 19 per cent in 1988 to 24 per cent in 1996. The import share also rose, though less sharply, from 16 per cent to 18 per cent over the same period.
4. The share of FDI stocks was lower at 4 per cent, reflecting the rapid build up of flows in the last decade.
5. These estimates are for the Republic of Korea, Malaysia, Thailand, Indonesia and the Philippines; hence they include data for Korea but exclude Singapore.
6. Driffield and Noor (1999) found that 90 per cent of the output of foreign-owned companies in the Malaysian electronics and electrical industry was exported. Firms were attracted to Malaysia by a combination of export incentives, low wages, and tax and investment incentives. EU firms were found to have few linkages with locally-owned enterprises.

7. See, for example, the Trade Facilitation Action Plan 1998 at: http://europea.eu.int/en/comm/dg01/0213asem.htm
8. Classified at the NACE 3-digit level.
9. Of the 90 industries with an RCA index of >0.001 or <-0.001 in 1988, 63 had less extreme values in 1996, including 8 which changed sign.
10. Das (1998) found that for their trade with all partners, Indonesia, Malaysia, Philippines and Thailand had a revealed comparative advantage in labour-intensive industries in 1993.
11. The data are from EUROSTAT (1992 and 1993), averaged for the years 1988–90.
12. The coefficient was significantly different from zero at the 1 per cent level for all years.
13. The GL index is sensitive to the level of aggregation at which the trade data are analysed, since the more disaggregated the data, the less scope there is for net exports of one component product to be offset against net imports of another component product, thereby increasing the recorded amount of intra-industry trade (the categorical aggregation problem). To be meaningful, the level of aggregation should be one at which the resultant industry groupings are relatively homogeneous.
14. Calculated as the value (in ecu) of imports or exports of a particular product (measured at the 4-digit level of the Harmonised System of trade classification) divided by the quantity of imports in tonnes.
15. This general pattern does not hold in the case of the important electrical machinery industry where VLIIT involves the exchange of high quality ASEAN products for low or medium-quality EU products. However two thirds of IIT in this industry is VHIIT.
16. See the *Financial Times* (16 February 1998).
17. See the *Financial Times* (7 October 1998).

BIBLIOGRAPHY

Akamatsu, K. (1962) 'An Historical Pattern of Economic Growth in Developing Countries', *The Developing Economies*, 1, 17–31.

Ariff, M. (1995) 'The Prospects for an ASEAN Free Trade Area'. In S. Arndt and C. Milner (eds), *The World Economy: Global Trade Policy 1995* (Oxford: Blackwell).

Authukorala, P.-M. and Menon, J. (1997) 'AFTA and the Trade-Investment Nexus in ASEAN', *The World Economy* 20, 2, 159–74.

Choy, I.W.T. (1998) *The Changing Pattern of ASEAN–EU Trade: 1988–1996* (Bruges: College of Europe) (unpublished).

Commission of the European Communities (1994) *Towards a New Asia Strategy*, COM(94) 314 final (Luxembourg: Office for Official Publications of the European Communities).

Commission of the European Communities (1997) 'EU Trade with Asian ASEM', *European Economy*, 3, 55–83.

Das, D.K. (1998) 'Changing Comparative Advantage and the Changing Composition of Asian Exports', *The World Economy*, 21, 1, 121–40.

Davis, E.P. (1998) 'Was the Asian Crisis a One-off Event?' *Oxford University Press Economics Newsletter*, Autumn.

Diwan, I. and Hoekman, B. (1999) 'Competition, Complementarity and Contagion in East Asia', CEPR, Discussion Paper no. 2112.

Driffield, N.L. and Noor, A.B.H. (1999) 'The South East Asian Financial Crisis and FDI in Malaysia. How 'Embedded' is the Foreign Owned Sector?' Cardiff Business School, Working Paper WP 99–024.

EUROSTAT (1992) *Structure and Activity of Industry 1988/89* (Luxembourg: Office for Official Publications of the European Communities).

EUROSTAT (1993) *Structure and Activity of Industry 1989/90* (Luxembourg: Office for Official Publications of the European Communities).

Greenaway, D., Hine, R. and Milner, C. (1994) 'Country-specific Factors and the Pattern of Horizontal and Vertical Intra-Industry Trade in the UK', *Weltwirtschaftliches Archiv*, 130, 152–74.

Greenaway, D., Hine, R. and Milner, C. (1995) 'Industry Characteristics in the Pattern of UK Vertical Intra-Industry Trade', *Economic Journal*, 105, 433, 1505–18.

Hill, H. and Phillips, P. (1993) 'Trade is a Two-way Exchange: Rising Import Penetration in East Asia's Export Economies', *The World Economy*, 16, 6, 687–97.

Imada, P.Y. (1993) 'Production and Trade Effects of an ASEAN Free Trade Area', *Developing Economies*, 31, 1, 3–23.

International Monetary Fund (1996) *Direction of Trade Statistics Yearbook* (Washington DC: IMF).

Messerlin, P.A. and Reed, G. (1995) 'Anti-Dumping Policies in the United States and the European Community', *Economic Journal*, 105, 433, 1565–75.

Morrissey, O. and Nelson, D. (1998) 'East Asian Economies – Miracle or Surprise?' *The World Economy*, 20, 1, 855–79.

Panagariya, A. (1994) 'East Asia and the New Regionalism in World Trade', *The World Economy*, 17, 6, 817–39.

Pomfret, R. (1993) 'Measuring the Effects of Economic Integration on Third Countries: a Comment on Kreinin and Plummer', *World Development*, 21, 3, 1435–37.

Schmitt-Rink, G. and Lilienbecker, T. (1991) 'An Analysis of ASEAN–EC Trade in Textiles and Electronics, 1980–88'. In N. Wagner (ed.), *ASEAN and the EC: the Impact of 1992* (Singapore: Institute of Southeast Asian Studies).

World Bank (1993) *The East Asian Miracle* (New York: Oxford University Press).

World Trade Organization (1998) 'Special Report: World Trade Growth Accelerated in 1997, Despite Turmoil in Some Asian Financial Markets', *WTO Focus*, no. 28, March, 1–7.

2 Foreign Direct Investment in ASEAN: an Historical Perspective

Anthony Bende-Nabende

INTRODUCTION

Foreign direct investment (FDI) has become one of the major influences on international economic relations. It has achieved this influence not only directly, but also by being closely linked with international trade and other financial flows. The early stages of investment may for instance lead to an increase in imports of capital machinery followed by imports of inputs such as components and parts required for the initial production process. Later when the production process has taken off, then FDI may well lead to an increase in exports. However, there are times when FDI replaces international export trade such as when countries adopt restrictive import-substitution industrialization strategies.

During the period from the 1950s to the 1970s, many economists and politicians in developing countries had a negative perception of FDI. In contrast, the current orthodoxy is that inward FDI is beneficial to host economies in that it contributes to the economic growth of developing countries through its positive spillover effects;[1] and to the economic growth of developed countries through increased savings followed by increased investment and consumption, as a consequence of dividends being paid to the shareholders when the profits are repatriated to the home countries. The existence of FDI, therefore, has a significant influence on the economies of both host and home countries.

This chapter takes an historical look at FDI in the ASEAN-5,[2] and attempts to explain the declining EU share of the total investment flow with reference to the ASEAN-5 economies' industrialization strategies and to concomitant developments in Japan, the Asian

33

Newly Industrializing Economies (NIEs), and Europe. The chapter is structured as follows. The first section explores the industrialization strategies of the ASEAN-5 economies during the colonial era, and explains the shift from an emphasis on primary production to import-substitution. The following section continues the story through the post-independence era, and the shift in development strategy from import-substitution to export-oriented manufacturing. Attention is also drawn to how production in the ASEAN economies was diversified towards non-traditional manufacturing activities, and to the liberalization of the regimes regarding foreign direct investment. The next section examines both the changes in the sectoral distribution of FDI in each of the ASEAN-5 economies, and changes in the provenance of this FDI. This statistical analysis provides the background for the discussion in the penultimate section of the reasons for, and the effects of, regionalism in Asia. The final section pulls together the general conclusions, and speculates briefly on future developments.

ASEAN INDUSTRIALIZATION STRATEGIES DURING THE COLONIAL ERA

In the colonial era, FDI in the ASEAN-4[3] was concentrated in raw material and resource-based manufacturing. The exception to this broad generalisation was FDI in Singapore, which was directed largely to manufacturing, and to commerce, finance and transport (which supported and promoted its entrepôt role), since it lacked natural resources and an agricultural base. The FDI was characterized by the foreign firms' oligopolistic activities, their pursuance of first-mover strategies to gain control over sources of raw materials, and their denial of these sources to their competitors. The firms were successful because, whereas the ASEAN-4 were in possession of abundant reserves of raw materials and Singapore had opportunities for the development of its manufacturing and services sectors, all five had neither the technology nor the skills necessary for the extraction and/or processing of raw materials, and for the development of the manufacturing and services sectors respectively. The foreign companies took advantage of this shortfall and provided both the technology and skills, enabling the raw materials in the ASEAN-4 to be extracted for export or for further processing and sale at both home and abroad, and enabling the development of the manufacturing and services sectors in Singapore.

This was an era when international trade was regarded as a major determinant of the economic growth of countries. Resource-endowed developing countries thus saw the trade in raw materials and processed goods as an engine of growth.[4] Trade between the producers and consumers of raw materials was promoted by the favourable demand conditions in the developed countries, not to mention the scarcity of raw materials in their economies combined with the desire to further their industrialization processes. It was further strengthened by the political–economic ties between the firms' home countries (colonial masters) and the host countries (colonies). The colonial masters dominated the trade of their colonies. For instance, the figures presented in Table 2.1 show that the Netherlands was responsible for 20.9 and 23.4 per cent of Indonesian trade in 1938 and 1950 respectively; that the United Kingdom accounted for 15.5 and 16.2 per cent of Malaysian trade in the same years; and that the United States dominated with 73.1 and 73.7 per cent of Philippine trade respectively.

During the colonial era then, the raw material and resource-seeking FDI in the ASEAN-5 was based upon political–economic links and was dominated by the colonial masters. It was driven by the respective countries' desire to grow economically, and was facilitated on the one hand by the abundance of raw materials and the lack of technology and know-how in the ASEAN-5 developing economies, and on the other hand by the favourable demand conditions combined with the availability of technology and know-how in the developed countries. Their trade links remained strong through the reciprocal trade of raw materials and finished products.

ASEAN INDUSTRIALIZATION STRATEGIES DURING THE POST-INDEPENDENCE ERA

The post-colonial era has been characterized by three interrelated developments which have affected both the volume and the nature of inward investment in the ASEAN economies: a change of development strategy from import-substitution to export-orientation,

Table 2.1   The Main Trading Partners of Five ASEAN Countries, 1938–95

ASEAN Country Trading Partner	Percentage of Total Trade of ASEAN Country										
	1938	1950	1955	1960	1967	1970	1975	1980	1985	1990	1995
INDONESIA											
European Union	20.9[1]	23.4[1]	13.6[1]	1.7[1]	26.3	19.1	12.7	10.1	11.9	15.3	17.5
United States	11.9	18.3	16.4	19.3	15.2	15.8	20.1	16.6	19.3	12.4	12.8
Japan	9.0	5.8	10.7	10.1	22.4	31.4	37.5	40.4	36.0	33.5	24.9
Asian NIEs	1.7	2.9	13.1	2.3	7.2	13.0	9.2	12.1	12.2	11.4	13.7
MALAYSIA											
European Union	15.5[2]	16.2[2]	9.5[2]	13.5	10.7	20.2	21.9	16.7	14.4	14.1	14.9
United States	4.6	16.2	1.4	7.2	11.7	9.2	13.4	15.7	14.1	17.3	18.7
Japan	3.0	5.6	5.8	10.3	14.3	12.9	17.3	22.9	23.5	20.6	20.4
Asian NIEs	4.1	1.1	25.0	2.0	16.5	16.8	15.8	19.0	23.3	24.0	22.4
PHILIPPINES											
European Union	n.a.	n.a.	n.a.	n.a.	10.6	12.0	14.3	14.1	12.1	14.4	13.8
United States	73.1	73.7	62.3	46.5	39.2	35.5	26.0	25.6	30.6	28.7	29.1
Japan	8.0	5.4	11.5	24.9	31.5	35.0	32.7	23.3	16.7	19.1	19.0
Asian NIEs	1.3	1.6	1.2	1.5	2.5	3.1	2.1	6.5	9.3	9.9	14.3
SINGAPORE											
European Union	n.a.	n.a.	11.3[2]	8.6[2]	7.1	8.6	13.5	11.9	11.0	13.7	13.4
United States	n.a.	n.a.	6.2	5.4	6.3	10.9	14.8	13.4	18.2	18.7	16.7
Japan	n.a.	n.a.	5.8	5.9	8.5	13.5	12.8	13.0	13.2	14.4	14.5
Asian NIEs	n.a.	n.a.	2.1	2.3	3.4	3.9	3.7	5.4	5.0	6.9	9.5
THAILAND											
European Union	7.1	6.8	n.a.	n.a.	13.9	20.4	16.6	19.5	17.0	18.0	15.5
United States	8.6	16.1	24.9	15.5	15.5	14.2	12.5	14.7	15.6	16.8	15.0
Japan	8.3	7.2	18.5	21.9	29.0	31.5	29.7	17.9	20.0	23.9	23.8
Asian NIEs	5.5	12.4	19.1	17.0	9.2	8.8	8.6	10.6	10.7	11.3	14.3

Notes: (1) Trade with the Netherlands only.
         (2) Trade with the United Kingdom only.
Source: OECD, Monthly Statistics of Foreign Trade (various years).

production diversification towards non-traditional manufacturing activities, and liberalization of the regimes regarding inward investment. These will be discussed below.

Change of Development Strategy

After attaining their independence,[5] the ASEAN economies started pursuing economic development based upon protectionist import-substitution strategies as the most immediate way of promoting their manufacturing industries.[6] These policies induced foreign investors to initiate 'tariff-jumping' investment that used overseas production to enter the domestic markets.

This early post-colonial industrial development came at a time when some politicians and economists from the developing countries associated FDI, and hence Transnational Corporations (TNCs), with negative perceptions such as acting as imperialist agents, using obsolete technology, practising unfair competition against local companies, using transfer pricing, and extracting excess profits.[7] Many developing countries thus instigated restrictive policy measures, particularly with regard to the ownership of natural resources.[8]

Despite these governments' attitudes towards foreign companies, most firms that had been established in these economies during the colonial era, and had survived the early post-colonial radical changes of host government policies, continued to flourish. This was achieved because, even if the indigenous small and medium-sized enterprises (SMEs) were being promoted and supported by their respective governments, the foreign firms had substantial ownership advantages over the indigenous firms that outweighed their disadvantage from being foreign. These advantages arose from their possession of advanced production technology, managerial resources, and marketing techniques. The continuity of the political–economic links established during the colonial era also made it possible for new firms from the former colonial masters to pursue the tariff-jumping FDI strategy, enabling them to access the market and produce behind the tariff walls. However, the newly-independent countries also started initiating bilateral trade and investment arrangements with economies other than their former colonial masters. As a consequence, the colonial masters gradually started to relinquish their dominance with regard to inward FDI.

The trade links, although remaining strong since the host countries relied on capital equipment, parts and components from their colonial masters, similarly started to decline in importance. For example, the Netherlands started losing share to the United States in Indonesia, the United Kingdom lost share to Japan in Malaysia, Japan started catching up with the United States in the Philippines, and both Japan and the United States caught up with the United Kingdom in Singapore.

Import-substitution initially provided the opportunity for rapid growth, above the rate of growth of domestic demand. After some time, however, several factors rendered this strategy non-viable. For instance, the domestic markets became saturated with locally-produced goods, while the heavy import demands (in terms of capital goods and industrial raw materials) of the light industries caused serious balance of payments problems. The recession of 1984–85 and its associated effects (i.e. declining external demand, large government budgets, balance of payments deficits, falling domestic savings, and increasing unemployment) acted as a big push factor for the ASEAN-4 to decisively revoke import-substitution and replace it with development strategies based on export-oriented manufacturing policies which also embraced diversification into non-traditional exports. It was implemented by the ASEAN-4 noticeably in the mid-1980s, although the policies had been gradually introduced since the 1970s.[9] In Singapore where such a policy had been implemented in the mid-1960s, several other factors were involved, including withdrawal from the Malay Federation, withdrawal of the British military forces and the consequent impact upon the economy, a small domestic market, and the lack of an agricultural sector.[10]

This new outlook encouraged investment geared towards export production, and hence an opportunity to focus on the larger external market where scale economies could be realized. The initial years comprised export-platform investment which was associated with the host countries' individual sources of comparative advantage, particularly their relatively lower wage costs. The emphasis was, therefore, put on the production of complete goods that required the TNCs to concentrate the production of labour- and resource-intensive goods in these economies, while keeping the production of capital- and/or technology-intensive goods in their home, or other capital-intensive, countries. This 'FDI rationalization strategy' was driven by the TNCs' desire to maximize their efficiency of production, and hence profitability, and was facilitated by the fact that the ASEAN-5 economies wanted to solve their then rampant unemployment problems. The ASEAN-5 economies later orchestrated a change in the type of investment away from the then traditional complete unit production, to sourcing and assembly.[11]

This export-oriented manufacturing encouraged specialization, and hence the regional division of labour in the Asian region as a whole and in the ASEAN-5 economies in particular, and encouraged further FDI rationalization by the foreign TNCs. It also promoted the growth of the SMEs through the associated linkages.

The role of the primary sector in the ASEAN-5 was consequently compromised by the development of the manufacturing sector. The growing importance of the services sector, on the other hand, was facilitated by the growth in demand resulting from increases in *per capita* incomes, technological advances in services provision, increasing liberalization, and the (traditionally) non-tradable nature of services FDI – though this is now changing (United Nations, 1997).

Production Diversification

The past two (three for Singapore) decades have also involved significant production diversification towards non-traditional manufacturing activities in the ASEAN-5 economies. Indonesia diversified into non-oil exports such as textiles and clothing, wood products (especially plywood), metals and machinery, foodstuffs and beverages processing, and footwear, all of which accounted for almost 50 per cent of the non-oil exports and 50 per cent of the growth in such exports between 1988 and 1992. Malaysia shifted its emphasis to labour- and resource-based exports, notably electronic components, textiles and clothing, and wood products. This was later extended to intermediate goods (e.g. chemicals) and, to a small extent, capital goods such as electrical machinery and motor vehicles. The Philippines shifted into products such as garments and electronic components, while Singapore put emphasis on labour-intensive manufacturing such as textiles and clothing, electronics assembly, furniture, petroleum refining, and rubber and wood processing. Singapore has been transformed into a modern manufacturing and financial centre supplying world markets, but the entrepôt trade in agricultural products remains important to its economy. In Thailand, diversification was made in the early 1970s to increase the variety of crops to include cassava, sugarcane, and pineapple together with a move towards manufacturing, notably of textiles and garments. Textiles have been the largest single export item since 1985, although there is strong competition from tourism (see Chapter 7) and electronics. Thailand also diversified into traded services.

These diversification policies created opportunities for the indigenous SMEs, that had been encouraged and supported by their respective governments during the period of import-substitution industrialization, to acquire strategic assets enabling them to operate in market niches in which they could not be threatened by the large and already well established Western TNCs. However, it also equally broadened the activities in which the foreign TNCs could invest and consequently justified further FDI rationalization.

Liberalization of Inward Investment

One of the most significant factors favouring the inflow of FDI into the ASEAN-5 economies has been the liberalization of their FDI regimes, concomitant with the diversification and export-oriented manufacturing policies discussed above. This liberalization was a major about-turn from the negative perceptions of FDI that pertained during the 1950–70s, and reflected a growing realization of the positive spillover effects of FDI through technology transfer, employment, human skills development, capital formation, and international trade.[12] It came at a time when it had become evident that most of the highly-protected and/or government-run industries/ projects were highly inefficient and unprofitable. The ASEAN economies responded by introducing new investment laws from the mid-1980s (the mid-1960s for Singapore) onwards. Encouraged by the conducive environment, FDI started flowing in, and more significantly so from non-traditional sources. As a consequence, international trade similarly shifted to non-traditional partners.

Each country has had a unique history of FDI liberalization, and these will be briefly outlined below. Singapore's policy towards foreign investment has been fairly liberal and non-interventionist ever since 1965. For instance, there are no limitations on equity ownership, no performance requirements, no foreign exchange controls/

limits, and no technology transfer requirements. It is only investments seeking incentives from the Economic Development Board (EDB) that are subject to certain requirements in terms of capital investment, level of technology, and the establishment of R&D facilities. The EDB administers tax incentives under the Economic Expansion Incentive (relief from Income Tax) Act 1959, which has been amended by the introduction of additional incentives in 1986 to make it more attractive. Incentives are now granted on a case-by-case basis and can be tailored to particular firms, providing a high degree of flexibility. The State, however, invests in industries which it thinks are crucial to its restructuring strategy. For example, the Operational Headquarters Incentive, introduced in 1986, was aimed at promoting companies to design, develop, produce, market, export and service their products from a Singapore-based operational headquarters.

In Indonesia, the government had taken steps since mid-1967 to encourage domestic and foreign investment in order to stimulate economic activities and growth. It did this partly through the enactment of the Foreign Investment Law in 1967, and the Domestic Investment Law in 1968. The Foreign Investment Law offered financial incentives and assurances for FDI, both for wholly-owned subsidiaries and for joint ventures with local firms. It also authorized the return of most of the Western enterprises placed under government supervision in 1963–64 to their former owners, except the Dutch whose enterprises had been nationalized and for which compensation had been negotiated. The Law was an about-turn from the nationalization policies of the 1950s and early 1960s but lasted only about five years, as the political protests of 1974 forced the government to focus on small-scale industries and solving the unemployment problem, and to become more restrictive and selective to FDI. For example, all new investments were to be joint ventures, Indonesian equity was to increase to 51 per cent within ten years, the list of closed industries was extended, and incentives were reduced. These policies were again reversed under the government Investment Priority List of 1977 but, shortly after this, Indonesia's oil revenue boomed in 1979 and the foreign investment requirements were tightened once more. The economic decline of the early 1980s prompted Indonesia to liberalize significantly its foreign investment regime, this time by dismantling many regulatory and approval controls from the 1960s and 1970s. The repatriation of capital and profits was thereafter allowed; the positive list of sectors open to investment was reduced to a negative list of closed or restricted sectors in 1989; and since 1992, profits have been allowed to be reinvested in foreign firms located in Indonesia and foreign exchange transfers have been unrestricted. The government, however, retains supervision of 'Strategic Commodity Industries' through the Strategic Management Agency established in 1989. Many of these industries are controlled by public enterprises, which are sheltered from competition by the government and which are characterized by poor performance and inefficiency.

FDI has been welcome in Malaysia since 1968, but aggressive pursuance only began in 1986 when the Promotion of Investment Act 1986 replaced the Investment Incentives Act of 1968. Earlier, the government had taken an initiative in 1970 to take over control of the key industries through mergers and acquisitions, especially in the mining and plantation industries, to the extent that 100 per cent foreign equity was only permitted in export-oriented industries, particularly electronics. The government's participation in the private sector was further increased through the institution of the Industrial Co-ordination Act in 1975, which required that all firms beyond a certain size register and comply with Ministry of Trade and Industry regulations. The new Investment Act provided for exemption from income and development tax for companies engaged in manufacturing new products or undertaking modernization, expansion and/or diversification. Consideration was to be given to a project's contribution to Malaysia's economic development, and to national and strategic requirements. There were to be special tax incentives for small scale companies, and for complying with the New Economic Policy guidelines; the 30 per cent *bumiputera* equity rule was to be flexible and hence negotiable; and the Malaysian Industrial Development Authority has been set up to assist foreign investors in matters such as finding local partners (Lim and Pang, 1991). The government reduced its key industries list to agricultural production and processing, forestry and forest products, manufacture of rubber products, manufacture of palm kernel oil products and derivatives, chemicals and pharmaceuticals, wood and wood products, textiles and textile products, electrical and electronic products, components and parts, and the manufacture of hand tools (Lim and Pang, 1991). Under the government's privatization plan, however, a number of

State-owned enterprises and projects were wholly or partly privatized between 1991 and 1992. In spite of these incentives, foreign investors still regarded the rules as unfavourable, particularly since the government tended to be actively involved at almost every level. For instance, despite government assurance that all new foreign investments made between 1986 and 1990 would be exempted from the NEP requirements, related issues such as the regulations governing new expansions made after 1990 on plants built between 1986 and 1990 remained largely unanswered. The government has, however, slowly loosened its grip on intervention and/or involvement to reduce unnecessary time-consuming bureaucracy.

There was no well-defined policy on foreign investment in the Philippines until an Investment Incentive Act was passed in 1968, followed by the Foreign Business Regulation Act in 1969, and an Export Incentive Act in 1970. There were also moves towards making the economy more responsive to international signals through the progressive removal of import restrictions from 1981 onwards. For instance, the Import Liberalization Programme initiated in 1981 was designed to improve the efficiency of resource allocation and to foster the development of internationally-competitive industries. An Omnibus Incentives Code was introduced in 1987, followed by the Foreign Investment Act (FIA) in 1991. The FIA expanded the number of economic sectors open to 100 per cent foreign ownership, streamlined the investment approval process, and more clearly defined the limits and restrictions on foreign investment (e.g. by establishing a short negative list of sectors where foreign investment continued to be limited). A major reform also involved the simplification of the tariff structure, whose average value came down from over 40 per cent to 25.6 per cent, with a further reduction to 20 per cent planned for 1995. Under a Presidential decree of August 1992, exporters were thereafter permitted to retain 100 per cent of their foreign exchange earnings. Several incentives, including tax and duty exemptions and tax credits, were offered under this programme. However, the government still retains control of the industries that it sees as strategic.[13] The restrictions written into the 1987 constitution also continue to limit foreign ownership to minority shareholdings in natural resource-related sectors.

In its encouragement of foreign investment, Thailand makes no distinction between local and foreign firms, except by specifying under the Alien Business Law and the Alien Occupation Law the sectors and occupations in which foreigners can engage. The government only screens foreign investment proposals that seek promotional privileges from the Board of Investment: special consideration is given to projects that contribute to the balance of payments, resource and regional development, energy conservation, linkage creation, employment, and the transfer of technology. Majority foreign ownership is permitted for export-oriented firms, but majority local ownership is required for firms producing for the domestic market: there is a 51 per cent local equity requirement in manufacturing for the domestic market, and a 60 per cent local equity requirement for projects in agriculture, animal husbandry, fisheries, mineral exploration and mining, and services (Lim and Pang, 1991). The tax incentives include *inter alia* exemption from corporate income tax, exemption or reduction from import duties, and exemption or reduction from business taxes on imported machinery and equipment, raw materials and components. Firms willing to locate outside the overcrowded Bangkok Metropolitan Area get additional incentives.

Comments

The early post-colonial import-substitution era in the ASEAN-5 was more or less similar to the colonial era, although with some shift towards more diversified FDI and trade links. However, the change of focus to export-oriented manufacturing, with its implications for scale economies and for the division of labour in the late 1970s and the 1980s, the diversification of production and exports into non-traditional sectors, and the liberalization of the FDI regimes decisively compromised the role of the primary sectors and increased the opportunity for FDI inflows by broadening the range of production activities in the manufacturing and services sectors in which foreign firms could participate. It also changed the nature of inward investment: from the raw material-seeking and resource-seeking FDI of the colonial era, to the tariff-jumping FDI of the early post-colonial import-substitution era, to production rationalization through FDI. Furthermore, competition from non-traditional FDI sources was stimulated, and the shares of both direct investment and trade in the ASEAN-5 started to change to the disfavour of the European Union.

TRENDS OF FOREIGN DIRECT INVESTMENT IN ASEAN

Figure 2.1 illustrates the sectoral distribution of the inward direct investment inflows for each of the ASEAN-5 economies since the early 1970s. The data relate to the percentage shares of the investment directed towards the primary, secondary, and tertiary sectors respectively and, despite the inevitable volatility associated with annual flow data, some interesting patterns emerge.

Figure 2.1   The Sectoral Distribution of Foreign Direct Investment in Five ASEAN Countries

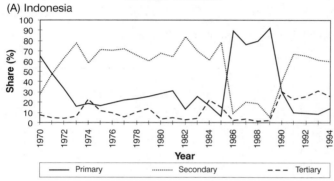

(A) Indonesia

Note: Data are for implemented FDI.
Source: United Nations, Transnational Corporations and Management Division (1992), UNCTAD (1997).

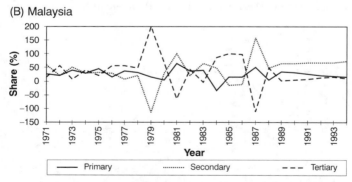

(B) Malaysia

Source: United Nations, Transnational Corporations and Management Division (1992), UNCTAD (1997).

Figure 2.1   (*cont.*)

(C) The Philippines

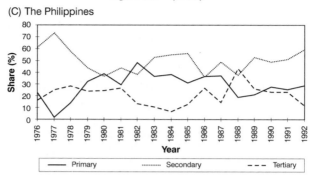

Source: Philippines' Board of Investment.

(D) Singapore

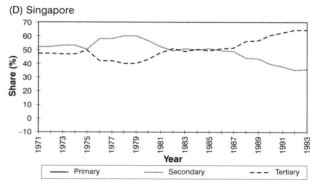

Source: Economic Development Board, Republic of Singapore.

(E) Thailand

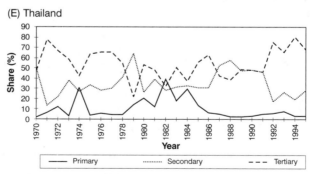

Source: Bank of Thailand.

For example, FDI in both Indonesia and the Philippines has been primarily in the secondary sector. In Indonesia, the primary sector dominated up to 1972, and again over the period 1986–89, but the secondary sector has otherwise accounted for over 60 per cent of total inward investment. Investment in the tertiary sector has been relatively insignificant, at least up to the 1990s. In the Philippines, the secondary sector has generally accounted for about 50 per cent of total inward investment, though the share of the primary sector was over 30 per cent between 1979 and 1987. The tertiary sector accounted for about 20 per cent throughout.

In contrast, the tertiary sector has played a much more important role with regard to inward investment in both Singapore and Thailand. In Singapore, FDI in the primary sector has been negligible, whilst the share of FDI in the tertiary sector has grown gradually at the expense of a gradually declining share directed to the secondary sector. In Thailand, the FDI inflows have been rather more volatile, but the importance of the tertiary sector has been evident throughout. Investment in the primary sector has been relatively small, except during the early 1980s when natural gas was discovered and developed. Finally, investment in Malaysia has exhibited no clear pattern, and is further complicated by the fact that negative figures are reported in several years. What does emerge, however, is that there has been considerable investment in all three sectors of the Malaysian economy.

Figure 2.2 highlights the provenance of the inward direct investment inflows for each of the ASEAN-5 economies since the 1970s/1980s. The data relate to the percentage shares of the investment provided by the four major investing countries/regions, namely: the European Union, the United States, Japan, and the Asian NIEs. Once again, despite the inevitable volatility associated with annual flow data, some interesting patterns emerge.

In Singapore, the European Union was the most important investor up until the early 1980s. Since then, the European Union, Japan, and the United States have all provided roughly equal shares, whilst the Asian NIEs' investment[14] has been limited to about 10 per cent of the total.

In contrast, the Asian NIEs have provided steadily increasing shares of total FDI flows to the other four ASEAN countries from the mid-1980s onwards. In Indonesia, the European Union accounted for a rising share of total FDI between 1978[15] and 1987 when it provided over 40 per cent, but this share had tailed off dramatically by the mid-1990s. A similar pattern is evident with the US data, whilst the Japanese share, notwithstanding some dramatic year-on-year fluctuations, has fallen steadily throughout the period. Meanwhile, the Asian NIEs' share had risen from about 10 per cent in the

Figure 2.2   The Provenance of Foreign Direct Investment in Five ASEAN Countries

(A) Indonesia

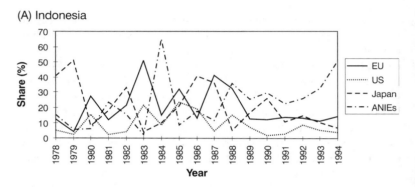

Source: Hill (1988), MITI (1995), United Nations, Transnational Corporations and Management Division (1992), UNCTAD (1997).

(B) Malaysia

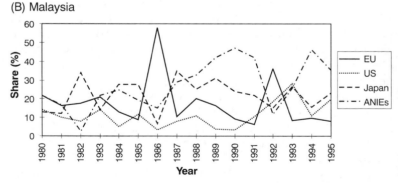

Source: MITI (1995), United Nations, Transnational Corporations and Management Division (1992), UNCTAD (1997).

(C) The Philippines

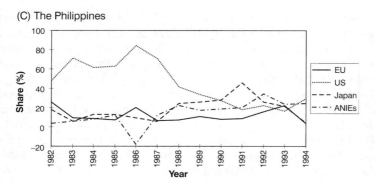

Source: Philippines Board of Investment.

(D) Singapore

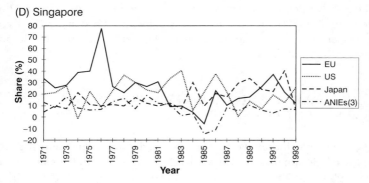

Source: Economic Development Board, Republic of Singapore.

(E) Thailand

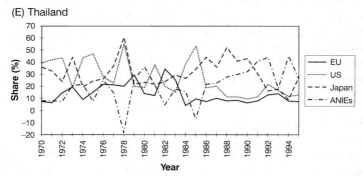

Source: Bank of Thailand.

mid-1980s to about 50 per cent by 1994. In Malaysia, the EU share has declined gradually since 1980[16] whilst the US share has increased somewhat. Japan has accounted for a fairly steady 20 per cent share, but again the most noticeable development is the increasing share being provided by the Asian NIEs, particularly after 1985 (ignoring a drop in 1992). In the Philippines, the United States dominated until 1990. Japan was an important investor in the early 1990s, but her share has fallen away whilst that of the Asian NIEs has risen steadily. The European Union has accounted for about 10–15 per cent of the total throughout. Finally, in Thailand, the EU share rose steadily through the 1970s, before peaking at about 30 per cent in 1982 and falling back to about 10 per cent of the total. Japan and the United States have been significant investors throughout, though their shares have declined from an average of about 40 per cent in the 1970s to about 20 per cent in 1995. Meanwhile the Asian NIEs have increased their share markedly since 1985, and accounted for 30 per cent of the total in 1995.

REGIONALISM, AND ITS IMPLICATIONS FOR INVESTMENT IN ASEAN

The data reported above reveal that the United States and the EU countries dominated FDI flows to the ASEAN countries up to the mid-1980s but that Japan, and more particularly the Asian NIEs, have since played a much greater role. These shifts in the relative importance of the various providers of direct investment reflect a growing regionalism in Southeast Asia which may be explained partly by the 'flying geese' model of economic development, partly by firms establishing regional networks and making greater use of regional sourcing and assembly operations, and partly by the enactment of various Bilateral Investment Treaties and Regional Economic Integration schemes. These developments will be considered in turn.

The Flying Geese Model

As countries develop, not only do their levels of income rise, and their technical skills and infrastructural competencies improve, but their factor costs increase as well. Eventually they lose comparative advantage in low-skill labour-intensive activities, and are forced to shift to higher value-added activities which require substantial investment in R&D, in strengthening the human resource base, and in infrastructural facilities.

Southeast Asia provides a good example of the relocation of production in accordance with this 'flying geese' model. During the 1970s and early 1980s, Japanese TNCs relied mainly on exports to service overseas markets, and most of the limited FDI was geared primarily towards supporting and maintaining the export strategy. Thus Japanese FDI was concentrated in financial services which supported the trading activities of its domestic firms in overseas markets, and in trade and distribution outlets in the United States and the European Union (United Nations, Transnational Corporations and Management Division, 1991). Since the mid-1980s, however, Japan has moved away from this concentration on export towards a greater reliance on FDI. As regards Japanese FDI in the Asian region, the most important reasons have been: (a) the increase in the value of the yen resulting from the Plaza Accord,[17] which made production within Japan increasingly less profitable; (b) the formation and extension of regional economic blocs, fears about the denial of access to the large markets of the European Union and North America, and thus the need for Japanese TNCs to create their own 'fortress' in Asia; (c) a surplus of funds as a result of Japan's high export earnings; and (d) the locational advantages offered by the Asian economies such as the liberalization of their FDI regimes, the availability of relatively cheap skilled labour, their improving infrastructural facilities, and the growth of their *per capita* incomes. The Asian countries in general, and the ASEAN-5 economies in particular, have therefore become important locations for Japanese firms, with regard to both production and consumption.

From the mid-1960s, the Asian NIEs had instituted policies[18] that promoted the growth of their indigenous SMEs. These policies combined with the dynamic economic growth of these economies stimulated the growth of their SMEs into well managed, efficient, technically competent, and professionally-operated enterprises. During the 1970s, while Japan retained the high-skill and high-technology activities such as R&D, the Asian NIEs shifted into the production of skill-intensive goods and services, as well as capital and technology-intensive items, while the Southeast Asian economies shifted out of resource-intensive production into the production of labour-intensive manufactures.

As the environment changed in the mid-1980s, the Asian NIEs' SMEs started diversifying their activities by investing in the relatively low-cost locations of the less-developed, neighbouring countries in Asia, in particular the ASEAN-4 economies, with which they shared both economic links and a familiar Chinese cultural background and influence. This investment was moreover facilitated by the opportunities that arose from the policies with regards to export-oriented manufacturing, production diversification, and liberalization, and from the general attractiveness of the ASEAN-4 economies.

The Asian NIEs now have large indigenous firms that possess advanced skills, research, and industrial bases, and invest heavily not only in electronics, automobiles, petrochemicals and oil refineries, but also in high value-added services. Over time, some have developed their own technological capabilities and have been transformed into globally competitive TNCs. They have developed their own intangible assets through R&D, acquired the ability to adopt or downscale mature technologies, and been prepared to exploit niche markets where many large TNCs do not wish to participate (United Nations Conference on Trade and Development, 1997).

In summary, the economic recession of the mid-1980s prompted the traditionally export-focused Japan and Asian NIEs to utilize their current account surpluses to rationalize their manufacturing operations, and to take advantage of the relatively low labour costs and culturally similar locations of the ASEAN countries. Companies from Japan and from the Asian NIEs began to capitalize on the opportunities that were available in the ASEAN economies at the time: i.e. the opportunity for the realization of scale economies as a result of the export-oriented manufacturing policies; the increased range of activities in which foreign companies could invest as a result of the production diversification policy; the more conducive environment for investment as a result of the liberalization policy; the increased potential for profitability as a result of the economic growth of the ASEAN economies; and the geographic proximity, cultural similarities, and economic links that reduced the 'economic distance'.[19] These factors significantly boosted Japanese investment in Singapore after 1985, and in the ASEAN-4 in the late 1980s and early 1990s, and equally boosted the Asian NIEs' investments in the

ASEAN-4. It similarly boosted trade links between Japan and the Asian NIEs, and the ASEAN economies. Since the mid-1980s then, Japan and the Asian NIEs have emerged as important investors in the Asian region. In addition to the reasons cited above, further stimulus has also been provided by capital liberalization in Taiwan, China and Korea, and the loss of the Asian NIEs' status as beneficiaries of the EU Generalized System of Preferences (GSP).

Regional Networks

Since the mid-1980s, companies from Japan and the Asian NIEs have been adopting strategies to pursue long-term growth by taking advantage of the corporate profits in their home countries, and the locational advantages of the Southeast Asian economies. They have partly achieved this by developing regional networks involving a production strategy of cross-border trade among affiliates to take advantage of scale and country specialization. The United Nations Transnational Corporations and Management Division (1991) has observed that these networks include affiliates in developing countries that are strongly linked to Japan. These affiliates are linked to markets in developed countries (for consumption purposes); serve local regional markets in countries where the affiliates are based (import-substitution); and act as low-cost suppliers to other affiliates located in Japan through intra-firm trade (rationalization), particularly in the electrical and electronic equipment, and automobile industries (United Nations, Transnational Corporations and Management Division, 1991). The value chain is thus broken into discrete functions, each being located wherever it can be carried out most efficiently and effectively, and whenever the penetration of an important growth market is facilitated by a reduction in the economic distance.

This strategic outlook has been facilitated by the different levels of economic development of the Asian countries, and from their comparative advantages which stem particularly from 'created assets': different qualities of human skills, of infrastructure, and of available technology. For example, Singapore currently acts as a regional procurement and operational headquarters, an R&D centre, an information and distribution centre (entrepôt role), and a site for the most sophisticated manufacturing operations. On the other hand, the ASEAN-4 countries have specialist capabilities in component manufacturing, and in resource-based and labour-intensive activities. Within the ASEAN-4 economies, there is also a division of labour according to the countries' relative comparative advantages: bottom-end work is typically located in Indonesia; while components manufacture and partial assembly are handled by Malaysia, the Philippines, and Thailand. The case of Toyota illustrates this clearly – see Figure 2.3. Singapore is used as the regional headquarters, while assembly is undertaken in the other four countries.

This strategy has allowed for plant specialization, and the exploitation of regional economies of scale by the TNCs, and has promoted intra-regional trade and investment in Asia. As similar networks

Figure 2.3   Toyota Automobile Networks in ASEAN

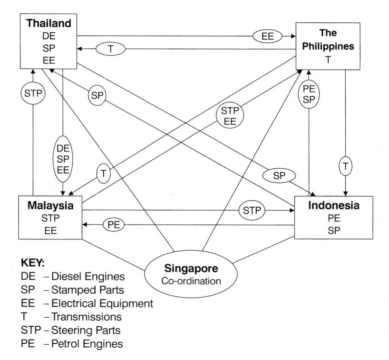

Adapted from United Nations, Transnational Corporations and Management Division (1991), p. 62.

have been implemented by EU and US firms in their respective regions, Japan and the Asian NIEs have strengthened their positions in the ASEAN countries and have had their FDI shares grow at the expense of the European Union.

Regional Agreements

Since the end of the 1980s, numerous regional Bilateral Investment Treaties (BITs) and Regional Economic Integration (REI) schemes have stimulated intra-regional trade and investment[20] around the world. Three major economic clusters – popularly known as the 'Triad' – have emerged centred on the European Union, Japan, and the United States. The main stimulants have been REI schemes in the European Union (the Single European Market programme) and US (the North American Free Trade Agreement) clusters, and regional BITs in the Japanese cluster.

With respect to FDI in the Asian region, the regional division of labour has led to new configurations. The ASEAN-4 are now clustered around the Asian NIEs, which in turn are clustered around Japan. Empirical studies on FDI flows into ASEAN provide mixed evidence on the link between the formation of ASEAN and AFTA, and changes in FDI inflows.[21] With respect to trade links, however, the ASEAN-5 are still clustered around Japan, though it is no doubt only a matter of time before the Asian NIEs form a trade-linked cluster as well.

In contrast, investment flows both into and within the European Union during the late 1980s were stimulated by the impending creation of the Single European Market (SEM). Anticipation of the SEM prompted an increase in cross-border mergers, acquisitions, and strategic alliances. Intra-EU FDI grew at a rate of 37 per cent per annum between 1980 and 1987, whilst inward FDI from other countries grew at only 17 per cent per annum (United Nations, Transnational Corporations and Management Division, 1991). The intra-EU FDI stock thus increased from 25 per cent of the total stock in 1980, to 40 per cent in 1988, and to over 50 per cent in 1990. Since 1989, moreover, more than 60 per cent of FDI from the then 12 Member States has remained within the European Union – see Figure 2.4. The United Nations, Transnational Corporations and Management Division (1991) has also observed significant amounts

Figure 2.4   The Destination of Outward Direct Investment from the
European Union

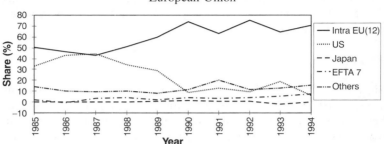

Source: OECD (1994), OECD (1995), Eurostat (various years).

of rationalized FDI by EU companies in Central and Eastern Europe, but these FDI flows declined in 1996 as a result of the slowdown in the privatization programme.

CONCLUDING REMARKS

FDI flows into the ASEAN-4 during the pre-industrial colonial and early post-colonial periods were largely associated with natural resources exploration and development, while flows into Singapore went largely into commerce, finance and transport. FDI was attracted to producing manufactures for the domestic markets of the ASEAN-5 in the 1960s and 1970s under the import-substituting industrialization strategies then in vogue. The political–economic links between the ASEAN-5 and their former colonial masters at the time meant that these countries dominated both FDI in, and trade with, the ASEAN-5.

The change in industrialization policies, combined with the liberalization of the FDI regimes, provided new opportunities for foreign investors: export-oriented manufacturing provided opportunities for the realization of scale economies, diversification broadened the activities in which foreign companies could participate, and liberalization offered a conducive environment and incentives for investment to be made. These opportunities became even more attractive with the realignment of major currencies in the mid-1980s.

Companies from both Japan and the Asian NIEs were quick to use FDI to rationalize their production activities throughout Asia. At the same time, EU companies were preoccupied with market-oriented FDI within the SEM, and with undertaking their own FDI rationalization strategies in Central and Eastern Europe. Together with the enactment of various treaties favouring regionalism, the result has been that Japan and the Asian NIEs have increased their respective shares of both trade and FDI in the ASEAN countries, whilst the relative importance of the European Union has correspondingly declined. As Europe, North America, and Asia continue their efforts to strengthen regional integration, the likelihood is that intra-regional investment and trade will dominate in the next decade. This is all the more likely if Asian integration is extended to encompass Japan, China, the Asian NIEs, and the ASEAN countries. If this happens, then the relative importance of EU–ASEAN trade and investment links may well diminish further.

NOTES

1. For empirical evidence based on econometric methods, see for example United Nations Transnational Corporations and Management Division (1993), Balasubramanyam *et al.* (1996), and Bende-Nabende *et al.* (1997a and 1997b).
2. The ASEAN-5 are Indonesia, Malaysia, the Philippines, Singapore and Thailand. The ASEAN-4 exclude Singapore.
3. Thailand was never a colony but, although it retained its political independence, its economy has relied heavily on the West, which directly or indirectly controlled the exploitation of and trade in its major commodities.
4. Petroleum and natural gas dominated Indonesia's exports, which also included primary and semi-processed agricultural and mineral commodities such as rubber, coffee, tin, shrimps, and palm oil. Malaysia's exports relied on the primary sector, and included resource- and agro-based products such as tin and rubber. The Philippines relied on the export of resource- and agro-based products such as copper ore and cathodes, coconut products, and lumber. Thailand's exports were dominated by agro-based products particularly rice and rubber, and other resource products such as tin and teak. Singapore exported manufactured goods and also served as a transit port for exports of primary goods from the other ASEAN countries.

5. The Philippines became independent in 1946, Indonesia in 1949, Malaysia in 1957, and Singapore in 1959.
6. Import-substitution, if well planned and well implemented, initially creates an industrial base and employment opportunities. In the 1960s, Malaysia protected its manufacturing sector particularly the industries for consumer goods such as food and household appliances; the Philippines protected its consumer goods industries particularly those producing processed food, textiles, pharmaceuticals, soaps, and detergents; Singapore's import-substitution industrialization policy was characterized by low tariffs on a narrow range of industries; and Thailand's import-substitution industrialization policy grew out of a programme of 'State Capitalism' undertaken during the 1950s when a number of state enterprises were set up to produce a wide range of manufactured goods. From mid-1967, the Indonesian government took measures to protect domestic industries through tariffs.
7. See, for instance, Streeten (1974) on the negative effects of FDI.
8. For example, in 1957, Dutch capital was expropriated and thereafter other foreign enterprises became nationalized by the Indonesian Government. In 1970, the Malaysian Government took over control of key industries through mergers and acquisitions, especially in the mining and plantation industries, to the extent that 100 per cent foreign equity was only available in export-oriented industries, particularly electronics. By the 1970s, most Thai industries were protected and in the hands of either local industrialists, or foreign (mostly Japanese)/local (mostly Chinese) joint ventures producing for the local market behind tariff walls.
9. For instance, in Malaysia the state agencies established Export Processing Zones (EPZs) where 100 per cent foreign ownership of export-oriented firms was permitted, and the state enterprises entered joint ventures with foreign firms. In the Philippines, EPZs were created under the authority of the Export Processing Zone Authorities, operating as self-contained enclaves for a variety of industries using imported raw materials and intermediate materials for processing, assembling and manufacturing of goods for export.
10. Singapore joined the Malay Federation in 1963 but was forced to quit in 1965. Faced with a challenge for survival after separation from Malaysia, it had to accelerate its export-oriented industrialization development, an issue that became even more urgent when the British military announced in 1967 that they were to withdraw by March 1971. The British forces accounted for 40 per cent of GDP and provided substantial employment, in addition to fulfilling a security role. Singapore consequently changed to an export-oriented manufacturing policy, with emphasis on labour-intensive manufacturing, aimed at solving the massive unemployment problem. This involved the liberalization of the FDI regime.

11. Sourcing and assembly involves capital- and technology-intensive parts/components being sourced in capital- or technology-rich countries, and the labour-intensive assembly process being undertaken in labour-rich countries. Sourcing typically requires the use of moderately skilled labour and moderately sophisticated machinery, while assembly utilises unskilled labour and unsophisticated machinery.

12. On the importance of FDI in this context see, for example, the International Labour Office (1977, 1981 and 1994), Hill (1993), United Nations Centre on Transnational Corporations (1992), Chudnovsky (1993), and the Organization for Economic Co-operation and Development (1995).

13. For instance, those industries perceived to be crucial to the accelerated industrialization of the country; that require large capital investment, highly specialized or advanced technology; that have strong backward and forward linkages; or that generate substantial foreign exchange savings.

14. Investment by Hong Kong, Taiwan, and the Republic of Korea.

15. Consistent data are only available for Indonesia since 1978.

16. The Malaysian data are only available from 1980, and refer to approvals rather than implemented FDI.

17. The Plaza Accord was reached by the Group of Five (G5 – France, West Germany, Japan, the United Kingdom, and the United States) in their September 1985 meeting at the Plaza Hotel, New York, and envisaged orderly appreciation of the major non-dollar currencies.

18. Under the import-substitution development strategies for the Republic of Korea and Taiwan, and the FDI liberalization policies for Hong Kong and Singapore.

19. 'Economic distance' is a concept that embraces information costs; political, historical, cultural, and economic links; cross-border communication costs; transport costs; and transaction costs.

20. An REI scheme creates locational advantages, so that higher investment takes place. For details, see United Nations, Transnational Corporations and Management Division (1993).

21. Tang's (1993) study has revealed that the APTA has not led to a higher level of investment and factor productivity growth among the member countries. Alun (1988) had earlier attributed the slow progress of ASEAN integration to the heterogeneity of its economies as a result of the differences in their economic growth rates, while Riedel (1991) argued that it is a result of the strict rule-of-origin restrictions, since each country maintains its own separate tariff schedule. Bende-Nabende *et al.* (1997a), however, found that ASEAN integration led to the diversion of investment from the less developed Indonesia and the Philippines, to the more developed Malaysia, Singapore and Thailand.

BIBLIOGRAPHY

Alun, J. (1988) 'ASEAN Production-Factor Mobility: Some Thoughts'. In H. Esmara (ed.), *ASEAN Co-operation: a New Perspective*, pp. 43–53 (Singapore: Chopmen).

Asian Development Bank (1994) *Key Indicators of Developing Asian and Pacific Countries 1994* (Singapore: Oxford University Press).

Balasubramanyam, V.N., Sapsford, D. and Salisu, M.A. (1994) 'Foreign Direct Investment and Growth', Lancaster University, The Management School, Discussion Paper.

Balasubramanyam, V.N., Sapsford, D. and Salisu, M.A. (1996) 'Foreign Direct Investment and Growth: New Hypotheses and Evidence', Lancaster University, The Management School, Discussion Paper EC7/96.

Bende-Nabende, A. (1999) *FDI, Regionalism, Government Policy and Endogenous Growth: a Comparative Study of the ASEAN-5 Economies, with Development Policy Implications for the Least Developed Countries* (Aldershot: Ashgate).

Bende-Nabende, A., Ford, J.L. and Slater, J.R. (1997a) 'The Impact of FDI and Regional Economic Integration on the Economic Growth of the ASEAN-5, 1970–1994: a Comparative Analysis from a Small Structural Model', University of Birmingham, Department of Economics, Discussion Paper no. 97–13.

Bende-Nabende, A., Ford, J.L. and Slater, J.R. (1997b) 'The Impact of FDI and Regional Economic Integration on the Economic Growth of the ASEAN-5, 1970–1994: A Comparative Dynamic Multiplier Analysis from a Small Model with Emphasis on Liberalisation', The University of Birmingham, Department of Economics, Discussion Paper no. 97–18.

Chen, E.K.Y. (1993) 'Foreign Direct Investment in East Asia', *Asian Development Review*, 2, 1, 24–59.

Chen, E.K.Y. and Wong, T.Y.C. (1995) 'Two-Way FDI Flow Between Hong Kong and Mainland China'. In Nomura Research Institute and Institute of Southeast Asian Studies, *The New Wave of Foreign Direct Investment In Asia*, pp. 243–77 (Singapore: Institute of Southeast Asian Studies).

Chia Siow Yue (1991) 'Foreign Direct Investment in ASEAN Economies', *Asian Development Review*, 2, 1, 60–102.

Chudnovsky, D. (1993) 'Introduction: Transnational Corporations and Industrialization'. In D. Chudnovsky (ed.), *Transnational Corporations and Industrialization*, pp. 1–28. The United Nations Library on Transnational Corporations, vol. 11 (London: Routledge).

Dunning, J.H. (1981) *International Production and the Multinational Enterprise* (London: George Allen & Unwin).

Enos, J.C. and Park, W.H. (1988) *The Adoption and Diffusion of Imported Technology: The Case of Korea* (London: Croom Helm).

EUROSTAT (various years), *European Foreign Direct Investment Yearbook* (Luxembourg: Office for Official Publications of the European Communities).

Franko, L.G. (1994) 'Trends in Direct Employment in Multinational Enterprises in Industrialised Countries'. In P. Enderwick (ed.), *Transnational Corporations and Human Resources*, pp. 87–118. The United Nations Library on Transnational Corporations, vol. 16 (London: Routledge).

Hill, H. (1988) *Foreign Investment and Industrialisation in Indonesia* (Singapore: Oxford University Press).

Hill, H. (1993) 'Foreign Investment and East Asian Economic Development'. In D. Chudnovsky (ed.), *Transnational Corporations and Industrialization*, pp. 163–216. The United Nations Library on Transnational Corporations, vol. 11 (London: Routledge).

Hood, N. and Young, S. (1993) 'TNCs and Economic Development in Host Countries'. In S. Lall (ed.), *Transnational Corporations and Economic Development*, pp. 76–105. The United Nations Library on Transnational Corporations, vol. 3 (London: Routledge).

International Labour Office (1977) *Social and Labour Practices of Multinational Enterprises in the Petroleum Industry* (Geneva: ILO).

International Labour Office (1981) *Multinationals' Training Practices and Development* (Geneva: ILO).

International Labour Office (1994) 'Multinationals' Training Practices and Development'. In P. Enderwick (ed.), *Transnational Corporations and Human Resources*, pp. 216–34. The United Nations Library on Transnational Corporations, vol. 16 (London: Routledge).

Lim, L.Y.C. and Fong, P.E. (1993) 'Vertical Linkages and Multinational Enterprises in Developing Countries'. In D. Chudnovsky (ed.), *Transnational Corporations and Industrialization*, pp. 82–99. The United Nations Library on Transnational Corporations, vol. 11 (London: Routledge).

Lim, L.Y.C. and Pang, P.F. (1991) *Foreign Direct Investment and Industrialisation in Malaysia, Singapore, Taiwan and Thailand* (Paris: OECD).

Ministry of International Trade and Industry, Malaysia (1995) *Malaysia International Trade and Industry Report* (Kuala Lumpur: MITI).

Organization for Economic Co-operation and Development (1994) *OECD Reviews on Foreign Direct Investment: Portugal* (Paris: OECD).

Organization for Economic Co-operation and Development (1995) *Foreign Direct Investment, Trade and Employment* (Paris: OECD).

Pina, C. (1994) 'Direct Employment Effects of MNEs in Developing Countries'. In P. Enderwick (ed.), *Transnational Corporations and Human Resources*, pp. 216–34. The United Nations Library on Transnational Corporations, vol. 16 (London: Routledge).

Rana, P.B. (1985) 'Foreign Direct Investment and Economic Growth in the Asian and Pacific Regions', *Asian Development Review*, 5, 1, 100–15.

Republic of Singapore (1996) *Singapore's Investment Abroad 1994–95* (Singapore: Department of Statistics).

Riedel, J. (1991) 'Intra-Asian Trade and Foreign Direct Investment', *Asian Development Review*, 9, 1, 111–46.

Streeten, P. (1974) 'The Theory of Development Policy'. In J.H. Dunning (ed.), *Economic Analysis and the Multinational Enterprise* (London: George Allen & Unwin).

Tang, D.C.O. (1993) 'An Econometric Study of Economic Integration Among the Asia-Pacific Economic Cooperation (APEC) Countries', Ph.D. Dissertation Abstract from the Internet.

UNCTAD (1993) *World Investment Report 1993: Transnational Corporations and Integrated International Production* (New York: United Nations).

UNCTAD (1997) *World Investment Report 1997: Transnational Corporations, Market Structure and Competitive Policy* (New York: United Nations).

United Nations Centre on Transnational Corporations (1992) *World Investment Directory 1992. Volume 1: Asia and the Pacific* (New York: United Nations).

United Nations, Transnational Corporations and Management Division (1991) *World Investment Report 1991: The Triad in Foreign Direct Investment* (New York: United Nations).

United Nations, Transnational Corporations and Management Division (1992) *World Investment Report 1992: Transnational Corporations as Engines of Growth* (New York: United Nations).

United Nations, Transnational Corporations and Management Division (1993) 'The Effects of Integration on the Activities of Transnational Corporations in the European Community: Theory and Empirical Tests'. In P. Robson (ed.), *Transnational Corporations and Regional Economic Integration*, pp. 99–123. The United Nations Library on Transnational Corporations, vol. 9 (London: Routledge).

World Bank (1995) *Global Economic Prospects and the Developing Countries* (Washington DC: The World Bank).

3 Foreign Direct Investment in ASEAN: a Contemporary Perspective

Hafiz Mirza, Axèle Giroud, Frank Bartels and Kee Hwee Wee

INTRODUCTION

Foreign direct investment (FDI) and transnational corporations (TNCs) have played an important role in the industrial development and economic growth of ASEAN countries. Together with the activities of domestic companies and national governments, FDI has been one of the principal motors for development throughout the ASEAN region. Between 1980 and 1997, for instance, the total stock of FDI rose ninefold from $25.2 bn to $227.5 bn – see Table 3.1. In particular, FDI has assisted in the industrialization and development of the entire region. For example, largely because of foreign investment, the share of manufacturing in Indonesia's exports rose from 13.7 per cent of total exports to 52.6 per cent between 1985 and 1996; for Malaysia the share rose from 31.7 per cent to 76.8 per cent; and similar spectacular increases took place across ASEAN, from Thailand to the Philippines (Giroud, 1999).

This chapter will compare the ASEAN strategies of TNCs from various countries, including TNCs from Europe, paying particular heed to the attractions of the region, the motivations of these TNCs, and their future investment intentions. The discussion will initially focus on the results of a major survey[1] (Mirza et al., 1997) conducted just *prior to* the Asian Financial Crisis, followed by an analysis of how strategies have changed since the crisis began.

However, the crisis is not the only reason for the changing stance of TNCs towards ASEAN. By the early 1990s, it was clear that although

Table 3.1   Inward FDI Stock in South and East Asia, 1980–97 (US$ million)

Region/Country	1980	1990	1995	1997
ASEAN	25242	95712	169848	227511
• Brunei	19	26	62	76
• Indonesia	10724	38883	50755	62147
• Laos	2	14	206	463
• Malaysia	6078	14117	38453	45203
• Myanmar	5	173	937	1113
• Philippines	1225	2098	6852	9931
• Singapore	6203	32355	55491	78062
• Thailand	981	7980	16775	23104
• Vietnam	7	66	397	7452
Other East Asia	8098	55588	180462	290144
• China	—	14135	128959	217341
• Japan	3270	9850	17831	33164
• Cambodia	Neg.	Neg.	307	1049
South Asia	2178	4252	10088	43111
• India	1177	1667	4528	11212
Total	35518	155552	360478	560766
ASEAN's share of the total (%)	71	61	47	41

Source: UNCTAD, *World Investment Report 1996* and *1998a*.

ASEAN has become a significant player on the world stage, certain region-wide policy decisions had to be taken if the Association's growth and development were to be sustained. The primary reason for concern was the declining share that ASEAN countries were receiving of FDI flows to the developing world, especially investment going to Asia – see Table 3.1. One of the decisions which were taken to arrest the apparent decline in TNC's interest in the region was the launch of the ASEAN Free Trade Area (AFTA) whose aim is to reduce tariffs to 0–5 per cent and eliminate non-tariff barriers to trade across the region by 2003. While AFTA has its analogues in other parts of the world, even bolder policies were to follow. In particular, at the 5th ASEAN Summit in 1995, the ASEAN Heads of Government declared their intention to create a regional arrangement for investment, a notion which subsequently became known as the ASEAN Investment Area (AIA). The AIA was inaugurated in 1998 and, by doing this, the Association has become a pioneer, for no regional grouping – whether in the industrialized or developing worlds – has yet created the fully-integrated investment-supporting environment that is envisaged by this agreement. This chapter will also look at the implications of AFTA and the AIA for TNC strategies – and for ASEAN economic development (ASEAN Secretariat, 1998).

THE INVESTMENT STRATEGIES OF TRANSNATIONAL CORPORATIONS IN ASEAN PRIOR TO THE ASIAN FINANCIAL CRISIS

The development of foreign direct investment in ASEAN has to be seen in the context of rising global FDI, as well of the proportion of this flowing to Asia. Between 1980 and 1997, for example, the stock of world FDI increased from about US$514 billion to around US$3,455 billion; within this, developing South and East Asia's share rose from 6.7 per cent of the total to 15.3 per cent (UNCTAD, 1998a). This was to some extent at the expense of other developing countries (at least proportionally) since the Asian share of FDI in developing countries also increased markedly from around 30 per cent of the total in 1980 to about 50 per cent in 1972.[2] The ASEAN countries have been among the largest recipients of FDI in Asia, although their collective stock of FDI has fallen to about 41 per cent of the total (from a high point of around 70 per cent) – see Table 3.1. China has been the most significant relative gainer, although India and some of the newly industrializing economies (NIEs) have also successfully increased their stock of FDI. Having said this, the *absolute* stock of FDI in ASEAN has risen over six-fold since 1980 and by two-thirds since 1990. This is no mean achievement.

Of course, a major reason for concern on the part of ASEAN members is the significant role that FDI has played in the capital formation and industrialization of some ASEAN nations – see Table 3.2 – and the possibility that this source may dry up. If these and *other countries* in ASEAN are to continue benefiting from inflows of transnational capital then they need to remain vigilant.

Table 3.2   The Importance of FDI Inflows in
Capital Formation in the ASEAN Countries, 1996

	FDI Inflows as a Proportion of Capital Formation (%)
Indonesia	8.5
Malaysia	11.1
Philippines	7.8
Singapore	27.5
Thailand	3.0

Source: UNCTAD (1998a).

It is also worth asking whether ASEAN should seek to encourage FDI from specific source countries or regions. The triad countries (the USA, the European Union and Japan) remain the largest investors on a global basis – see Table 3.3 – and certainly ASEAN should continue its promotion of FDI from these sources. It is worth bearing in mind that the European Union is now the world's largest foreign direct investor by far, although an appreciable share of the FDI is in the form of cross-investments between EU countries. In addition, however, it is generally recognized that Asian countries themselves are becoming significant foreign investors. Table 3.3 shows that East Asian investors such as Hong Kong, China, Taiwan and South Korea are now established international investors and should be seen as sources of FDI to ASEAN, especially in a regional context. A number of ASEAN countries, especially Singapore, Malaysia and Thailand, are also investing increasing amounts abroad (this FDI is both by local firms and by foreign subsidiaries based in these countries) and the issue of intra-ASEAN investment is therefore also to the fore.

European and US investment in ASEAN began to falter in the early 1990s – see Table 3.4 – and further effort to attract investors from these regions might be worthwhile. However, in the context of NAFTA and the Single European Market project (possibly reinvigorated because of the successful launch of the euro), it might be argued that such efforts may pay less dividend than further encouraging TNCs from Japan and developing Asia: the latter have been the primary investors in China, ASEAN, and other parts of Asia

Table 3.3   FDI Stocks by Source Country, 1980–97
(US$ billion)

Country/Region	1980	1985	1997
1. Industrialized Countries	453.0	581.3	3192.5
• USA	220.2	251.0	907.5
• European Union	213.2	286.3	1596.9[1]
• Japan	19.6	44.0	284.6
• Australia	2.3	30.1	52.4
2. Developing South and East Asian Countries	1.7	5.6	279.4
• Hong Kong	0.1	2.3	137.5
• South Korea	0.1	0.5	18.0
• China	—	0.1	20.4
• Taiwan	0.1	0.2	34.2
• Singapore	0.6	1.3	43.4
• Indonesia	—	—	4.2
• Malaysia	0.4	0.7	15.7
• Philippines	0.2	0.2	1.2
• Thailand	—	—	3.8
• Pakistan	—	0.1	0.3
• India	—	—	0.1
3. Other	59.5	92.5	69.5
4. World	514.2	679.4	3541.4

Notes: (1) Includes intra-EU investment.
(2) — denotes negligible or no investment.
Source: UNCTAD (1998a).

in recent years, although it is difficult to discern exact amounts.[3] For instance, the best estimates suggest that there is something of the order of US$2 bn of Korean FDI, US$7 bn of Taiwanese FDI, US$3 bn of Australian FDI, and US$28 bn of Hong Kong/PRC FDI in ASEAN.[4] Having said this, the alternative argument is that EU investors cannot be ignored, both because of their sheer scale and, more importantly, because trade and investment relations between the European Union and ASEAN are much weaker than those between ASEAN and its other principal partners (the USA, Japan and the Asian NIEs) – for further elucidation of this argument, see EU/UNCTAD (1996) and UNCTAD (1996). One independent basis for assessing the viability of an ASEAN campaign to attract EU FDI is an annual survey of TNC strategies by Arthur Andersen which,

Table 3.4   FDI Stock in ASEAN by Source Country, 1980–93

	1980		1985		1993	
	Amount (US$ bn)	*Share of Total FDI (%)*	*Amount (US$ bn)*	*Share of Total FDI (%)*	*Amount (US$ bn)*	*Share of Total FDI (%)*
EU	4.2	18.8	7.7	21.7	23.3	16.5
Japan	5.7	25.4	9.2	25.9	33.0	23.4
USA	2.8	12.5	6.2	17.5	18.5	13.2
Other	9.7	43.3	12.4	34.9	66.1	46.9
Total	22.4	100.0	35.5	100.0	140.9	100.0

Source: UNCTAD-DTCI, FDI Database.

among other issues, looks at their future locational strategies. It is interesting to note that, while the 1997 survey indicated that TNCs from North America and developing Asia were more likely to expand their international investments most rapidly in the period until the year 2000, the latest survey – conducted a year after the Asian financial crisis began – indicates that it is EU/West European firms which are now most aggressive in expanding to areas (China, ASEAN, see also Table 3.4) where they are currently under-represented (Arthur Andersen, 1997, 1998). Certainly ASEAN countries, and the region as a whole, might now be advised to look to Europe as a significant source of future investment.

RESULTS FROM THE BRADFORD ASEAN SURVEY

Based on a survey of 311 North American, Asian and European TNCs, the 1997 Arthur Andersen survey mentioned above concluded that the most important reason for future investments in all regions, including developing Asia, would be market access (the 1998 survey was similar, but there were nuances which will be discussed later). This orientation has to be taken into account by ASEAN whose diversity includes a large market (hence the importance of AFTA), a range of skills, both low and high wages, and meaningful resources in some countries. On the plus side, the survey (which was conducted prior to the Asian crisis) also reported that developing Asia had the highest priority for all investors (excluding investments in the home region): Asia's ranking was 2.7, followed by Western Europe (2.0), North America (2.0), Eastern Europe (1.4), Latin America (1.3), Japan (1.1), and Africa (0.3).

The Significance of ASEAN for Foreign Investors

The findings of the Bradford ASEAN FDI survey (for a description, see endnote 1) broadly concur with those of the Arthur Andersen survey. (The description of the survey results in the remainder of this section will not refer to the Asian financial crisis: the impact on corporate views will be examined in the following section. The latter section will also assess whether the crisis has created a complete breach with the past or whether the sort of objectives and orientations discerned by the survey will resume.) The vast majority of foreign executives believe that ASEAN will play a greater role in their firm's future international strategy – see Table 3.5: this view is shared, almost entirely, by companies from all the investing countries. Business growth and regional opportunities were the preponderant reasons cited by firms for this view, although Japanese parents also

Table 3.5   The Views of Foreign Parent Companies regarding the Future Significance of ASEAN

	No. of Responses		
	More Significant	*Less Significant*	*Reasons*
Japanese (52*)	49	3	Business Growth, Opportunities, Ideal Production Site
Taiwanese (34*)	23	1	None Given
Australian (13*)	13	0	Business Growth, Market Potential
USA (21*)	19	2	Business Growth, Opportunities
European (19*)	16	3	Business Growth, Opportunities

Note: * The figures indicate the numbers of firms responding to the question.
Source: Bradford ASEAN Foreign Direct Investment Database.

saw prospects in terms of using ASEAN as a production base. Other ASEAN countries will increasingly be a *market* for regionally produced goods, especially for Japanese companies. Table 3.6 indicates that Japanese TNCs are likely to integrate markets and operations in East Asia in the future (e.g. there will be more ASEAN sales in China and 'other Asia'); this trend is also discernible for European firms, but is not so clearcut. Taiwanese TNCs appear to have already partly integrated their ASEAN and Chinese operations, but there may be no further integration. US firms, on the other hand, appear to see Asian sub-regions in discrete terms, although they too will increasingly export to other East Asian markets. Interestingly, companies based in ASEAN will export proportionally less to Japan and Taiwan in the future: this probably reflects the rapid growth in the size of developing Asian (especially ASEAN) markets, rather than a decline in the Japanese or Taiwanese markets in absolute terms. There are clear implications for ASEAN to support TNC integration in their cross-Asia/trans-ASEAN operations.

In terms of ASEAN's standing versus other regions, TNCs from all triad countries tend to give a high priority to Asian regional markets – see Table 3.7. ASEAN is more highly prioritized by Japanese, Australian and European firms, while US companies lean a little towards 'Greater China'.[5] Given that 'Greater China' includes Taiwan, it is clear that ASEAN also figures highly in the priorities of Taiwanese firms. Both China and ASEAN are highly regarded regions, however, despite the increased corporate interest in South and Central Asia. The fear that China will suck away investment from ASEAN appears to be unfounded.

The Future Investment Intentions of Foreign Investors in ASEAN

At a country level, all the larger ASEAN countries are significantly prioritized by foreign investors – see Table 3.8 – albeit China and India receive very high individual scores because their sheer size captures the corporate imagination. In terms of definite plans and 'seeking opportunities', Brunei features little because of its size and a market which can be easily served by imports; Singapore is probably now less prioritised for *new* investments because of its existing success in attracting TNCs (it is the only ASEAN country whose

Table 3.6 The Main Markets for the Foreign Subsidiaries in 1997 and 2000[1]

Market[2]	Japanese (67*)		Taiwanese (33*)		Australian (9*)		USA (22*)		European (19*)	
	Today	2000	Today	2000	Today	2000	Today	2000	Today	2000
ASEAN	27	50	21	23	9	9	16	17	11	12
Japan	54	43	7	8	3	3	5	3	3	2
China	10	32	19	19	2	4	1	8	2	7
S. Korea, Taiwan	10	10	20	15	3	3	5	4	2	4
Other Asia	1	9	10	8	4	5	5	7	5	6
Australia, Oceania	—	—	4	3	2	3	1	3	4	5
North America	36	33	11	9	5	2	8	5	6	5
European Union	25	21	9	8	3	3	6	7	8	8
Latin America	3	2	4	4	2	—	3	2	1	2
Africa	—	—	2	2	—	—	—	1	2	1
Eastern Europe	—	3	1	2	1	1	1	3	2	3
Other	2	2	—	—	—	—	—	—	—	—

Notes: (1) The survey was carried out in 1997, so the figures for the year 2000 refer to expected sales.
(2) The figures refer to the numbers of firms citing the markets as receiving 20 per cent or more of the subsidiaries' output.
  * The figures indicate the numbers of firms responding to the question.
Source: Bradford ASEAN Foreign Direct Investment Database.

Table 3.7 The Priority Markets of the Foreign Parent Companies[1]

Region	Japanese (68*)		Taiwanese (31*)		Australian (13*)		USA (21*)		European (18*)	
	High Priority	Med. Priority	High Priority	Med. Priority	High Priority	Med. Priority	High Priority	Med. Priority	High Priority	Med. Priority
NAFTA	18	25	5	10	6	1	14	3	7	2
Central & South America	1	9	—	—	1	1	4	10	2	5
European Union	5	23	1	5	2	2	6	6	8	3
Eastern Europe	1	4	—	—	—	2	1	5	1	5
Middle East	1	7	1	—	—	2	1	4	1	1
South Africa	—	4	—	—	—	2	—	3	—	2
Other Africa	—	4	—	—	—	1	—	3	—	1
Greater China	44	18	27	3	6	5	13	7	6	9
ASEAN	55	10	23	8	9	1	6	8	9	6
Northeast Asia	11	15	1	4	1	2	4	4	2	2
South Asia	4	21	7	7	—	3	3	3	4	2
Central Asia	3	8	—	8	1	1	1	3	1	2

Notes: (1) High priority regions received rankings of '1' or '2'; medium priority regions received rankings of '3' or '4'.
      * The figures indicate the numbers of firms responding to the question.
Source: Bradford ASEAN Foreign Direct Investment Database.

inward FDI is predominantly from industrialized countries), although *re*-investments will most likely continue in a major way. US TNCs are possibly a little less focused on Asia and ASEAN than TNCs from the other countries in this survey. These priorities also translate into investment plans for investors, especially with regards to opportunities in Indonesia, Malaysia, the Philippines, Thailand and Vietnam, with a few localized nuances. Both Brunei and Singapore figure in corporate strategic thinking. Also of interest, is the fact that plans are being laid by some TNCs from all countries for investment in new ASEAN member countries (Laos and Myanmar), as well as potential members (Cambodia).

Bearing in mind that ASEAN as a region is significantly smaller than either China or India, it is worth noting that 65 per cent of all definite investment plans by the surveyed companies relate to ASEAN countries taken as a whole. This is considerably more than for China (23 per cent) or India (10 per cent); of course the size and quality of the FDI are also factors which need to be taken into account, but this is nevertheless a positive finding for ASEAN. The same picture emerges in terms of companies seeking opportunities or waiting for one to arise. Moreover, according to surveyed firms, this situation is likely to continue into the longer term. When companies were asked about their future priority markets, ASEAN countries were chosen more often than India or China. Furthermore, support for inward FDI should not simply be viewed in competitive terms: ASEAN, China and South Asia will in the future increasingly trade with each other. As part of broader Asian economic integration, cross-investments between these major sub-regions have already begun.

The Factors Motivating Foreign Investment in ASEAN

Why are foreign investors attracted to East Asia, especially ASEAN? According to the ASEAN survey – see Table 3.9 – companies generally find five factors most compelling in priority markets. In rank order, these are a large potential market, political and economic stability, high growth rates, a favourable business climate, and few barriers to entry.[6] In the main, these are seen to prevail in both ASEAN host countries and ASEAN as a whole. For example, two-thirds of companies believed that ASEAN has low-cost labour, which is quite a good proportion given that the host and region are being

Table 3.8 The Plans of the Foreign Parent Companies for Investing in ASEAN or Nearby Countries[1]

	Japanese (65*)				Taiwanese (26*)				Australian (13*)			
	Defn. plans	Seeking Opps.	If Opp. arises	No plans	Defn. plans	Seeking Opps.	If Opp. arises	No plans	Defn. plans	Seeking Opps.	If Opp. arises	No plans
Brunei	1	—	2	40	—	—	1	8	—	—	1	7
Indonesia	12	17	15	13	4	5	8	1	4	3	2	1
Malaysia	5	6	8	27	8	3	4	3	2	2	3	2
Philippines	8	8	17	18	6	—	5	3	3	2	2	1
Singapore	4	5	8	28	5	3	2	3	1	2	4	2
Thailand	18	5	12	16	7	2	3	3	3	2	3	2
Vietnam	7	14	28	16	5	11	9	2	3	1	6	1
Cambodia	—	2	11	34	—	—	3	5	—	—	4	5
Laos	—	1	9	35	1	—	3	4	—	—	4	5
Myanmar	1	3	16	30	—	—	3	5	—	2	3	4
China	20	19	14	9	10	10	5	1	4	3	1	2
India	4	18	19	20	1	2	1	6	2	3	1	3
Other	—	—	1	13	—	—	—	4	1	—	3	—
(ASEAN)	80	98	160	299	47	36	47	48	23	20	37	35

(continued)

Table 3.8 (continued)

	USA (22*)				European (19*)				Total			
	Defn. plans	Seeking Opps.	If Opp. arises	No plans	Defn. plans	Seeking Opps.	If Opp. arises	No plans	Defn. plans	Seeking Opps.	If Opp. arises	No plans
Brunei	—	1	4	13	—	—	2	11	1	1	10	79
Indonesia	12	6	3	1	4	4	4	2	36	35	32	18
Malaysia	4	10	5	2	9	3	—	3	28	24	20	37
Philippines	4	4	7	7	4	3	4	3	25	17	35	32
Singapore	5	5	4	4	4	2	—	4	19	17	18	37
Thailand	7	5	4	5	4	2	3	2	39	16	25	28
Vietnam	5	6	4	7	7	6	3	2	27	38	50	28
Cambodia	—	1	3	15	2	1	1	10	2	4	22	69
Laos	1	1	1	15	1	2	1	10	3	4	18	69
Myanmar	2	1	1	15	1	2	3	9	4	8	26	63
China	18	4	—	—	11	6	1	1	63	42	21	13
India	12	5	1	4	9	6	2	1	28	34	24	34
Other	1	—	—	—	1	—	—	3	3	—	4	20
(ASEAN)	71	49	37	88	57	35	24	61	278 (65%)**	238 (67%)**	305 (74%)**	531

Notes: (1) The figures give the numbers of responses from firms indicating investment plans over the following five years.

    * The figures indicate the numbers of firms responding to the question.

    ** The percentage figures are the ASEAN share of all responses.

Source: Bradford ASEAN Foreign Direct Investment Database.

Table 3.9 The Most Important Characteristics of Priority/Promising Markets for Foreign Investors[2]

Characteristics[1]	Generally	Host Country	ASEAN
Large Potential Market	99	42	74
Few Barriers to Entry	38	41	18
Political & Economic Stability	78	51	45
High Potential Rate of Return	36	20	21
High Growth Rates	53	34	43
Psychic/Cultural Proximity	1	5	4
Protection of Intellectual Property	11	8	5
Good Commercial/Legal System	21	15	14
Opportunities to Acquire Firms	5	3	1
Privatisation Programme	10	4	4
Favourable Business Climate	44	35	25
Strategic Business Location	21	13	19
Government Assistance	6	5	4
Tax Incentives	4	8	11
Low Cost Labour	34	23	23
Access to Raw Materials	15	11	8
Actions of Competitors	6	5	6

Notes: (1) The foreign investors were asked to indicate the four most important characteristics of a priority or promising market generally. They were then asked to indicate whether these characteristics obtained for their primary ASEAN host country, and for ASEAN as a whole. Their responses were not entirely consistent, but this was only a problem for a small number of respondents.

(2) There were 123 responses, from TNCs in all countries except Taiwan. The responses by the Taiwanese firms were excluded because of a possible mistranslation/misunderstanding of the question.

Source: Bradford ASEAN Foreign Direct Investment Database.

compared against an ideal. (Also this result again suggests that ASEAN should be less concerned about the 'China factor'.) There are some nuances, however:

- 74 of the 99 companies which consider market size important believe that ASEAN has a large market, but only 42 regard the same to be true about their host country – in other words ASEAN countries have to stick together in order to ensure that firms continue to target the region as a market.

- Similarly, 43 of the responses regard the ASEAN region as having high growth rates (this factor being considered as being important by 53 firms), but only 34 feel the same about their host country.
- In contrast, the host is regarded as being more stable than ASEAN as a whole (51 responses out of a possible 78, versus 45), perhaps because of greater familiarity with host country institutions; and 35 responses (out of a possible 44) indicate the perception that the host's business climate is favourable, whereas this figure is only 25 for ASEAN. In other words, companies are looking to ASEAN as the fundamental or ultimate target for their investments, but would like the familiarity of their host environment to be more generally available in ASEAN.

These results strongly testify to the need for ASEAN countries to grow together in terms of the ASEAN Free Trade Area – which is well on track – and the ASEAN Investment Area – which was recently inaugurated, but whose provisions are still being negotiated. Furthermore, these findings also suggest that ASEAN countries need to create a strong regional identity and institutions. A recent paper by the Japan Institute for Overseas Investment (JOI) indicates that Japanese firms have adjusted their regional (ASEAN) policies in response to the establishment of AFTA (JOI, 1997).

Finally, although government support and promotion measures are relatively less important, the high significance of the business climate in Table 3.9 shows that they *are* important. This can also be seen from Table 3.10, which looks more specifically at certain measures. Thus, for example, although 128 companies (out of 145) regarded the market as a major or minor reason for investing in ASEAN, a favourable business climate influenced no less than 113 of them in a major or minor way. Again, while political and economic stability was important, so were especially fiscal incentives, financial assistance and other privileges (for up to 91 firms out of 145 in terms of taxes and tariffs). The liberalization measures (on foreign exchange movement, profit repatriation) and infrastructural developments were also highly appreciated. A number of companies (mainly Japanese) were very familiar with very specific ASEAN schemes, such as brand-to-brand complementation (JOI, 1997). Government policies and measures are likely to be accorded more significance in the future.

Table 3.10   The Effects of Government Measures on Foreign Direct Investment in ASEAN Countries[2]

Reason[1]	Major Reason	Minor Reason	Likely Future Reason
Large Potential Market in Host Country	113	15	30
Large Potential ASEAN Market	79	34	38
Favourable Business Climate	55	58	20
High Quality Infrastructure	41	47	15
Political and Economic Stability	69	57	17
Liberal Foreign Exchange Regulations	34	51	17
Liberal Profit Repatriation Regulations	35	56	19
Strong Patent Protection	12	36	11
Financial Assistance from Govt.	18	29	8
Tax or Tariff Incentives	51	40	11
Exclusionary Licensing etc.	14	39	5
Govt. Schemes	3	21	4
ASEAN Schemes (inc.:)	6	24	7
• ASEAN Ind. Joint Ventures	3	10	6
• Brand-to-Brand Comp.	4	8	5
• Growth Triangles	7	7	7

Notes: (1) Companies were asked the major and minor reasons for their original investment in their primary ASEAN host country, and also whether the reason would apply in the future. The number of responses for the future are lower because only a certain number of new investments are likely. The table excludes reasons other than those items directly affected by government policies (except broad market reasons).
(2) There were 107 responses, from TNCs in all countries.
Source: Bradford ASEAN Foreign Direct Investment Database.

THE POST-CRISIS INVESTMENT STRATEGIES OF TRANSNATIONAL CORPORATIONS IN ASEAN

The influence of the crisis on foreign direct investment and transnational corporate activity in ASEAN can be discussed at three levels: (i) the impact on the scale of inward FDI, (ii) the restructuring of TNC operations, and (iii) changes in regulations and policies pertaining to FDI. Turning to the first issue, the level of inward FDI has been remarkably stable compared to foreign *portfolio* investment and international bank loans. According to UNCTAD, FDI in 1997 was only marginally down (about 5 per cent) compared to 1996 in the four most affected ASEAN countries (Indonesia, Malaysia, the Philippines, and Thailand) and was actually running at a higher rate in early 1998 (UNCTAD, 1998b: 2) – however the latter figures are extremely provisional and the difficulties in Indonesia had not yet escalated to violent levels.

Having said this, it is clear that there will be no sudden pull-out from ASEAN (or other parts of Asia) and, indeed, there are reasons why TNCs might find FDI more attractive. The reason for the latter is that the costs of doing business in ASEAN are now much lower than before: currencies have depreciated, property prices have plummeted, local assets are cheaper, and local companies (especially those in difficulty) have become available for acquisition. This has presented opportunities for cash-rich foreign companies, especially those from the United States, Europe, and Singapore, but also TNCs from Japan, Hong Kong, and Taiwan.[7] FDI in some industries in ASEAN, such as financial services and telecommunications, is growing more rapidly than before because of unexpected opportunities. European companies such as British Telecom, Volvo, BASF and Usinor-Sacilor have been involved in a spate of high-profile mergers and acquisitions. In overall terms, the 1998 Arthur Andersen survey reported that ASEAN/Pacific Asia (excluding China) was still the highest-ranking priority region for TNCs, despite the region's travails.

There is also some evidence that the structure of TNC operations in ASEAN is changing. This is especially apparent in a shift towards export orientation (because of increased international price competitiveness as a consequence of depreciated currencies) and a shift away from domestic market orientation (because of reduced local real incomes), especially in the four most affected countries (UNCTAD, 1998b). Having said this, in spite of the scale of intra-TNC trade, it will not be possible for companies to simply sell their produce in other markets: Japanese consumers are not buying, the European and North American markets have a finite capacity for many types of good being produced in ASEAN, and other developing countries are also trying to pursue an export-orientated path. It is therefore quite possible, provided there is no further economic volatility, that the crisis will have resulted in an acceleration in the development of ASEAN's regional division of labour in line with the aims of AFTA and the AIA (see below). Certainly an increase in the sale of goods to other markets in ASEAN, a spread-out of TNC activity across the region (building on existing regional production networks), and the rise of intra-ASEAN FDI (especially by Singapore) all testify that this eventuality is possible – and desirable.[8]

The strategies of TNCs towards ASEAN have also been affected by changes in regulations and policies. As mentioned earlier, mergers and acquisitions by foreign firms in ASEAN countries have increased since the start of the crisis in 1997 because desperate governments have relaxed their regulations in order to encourage the buy-out and turnaround of ailing indigenous firms. The Thai banking and financial sector is a case in point (UNCTAD, 1998b). While there is the danger of a 'fire-sale' of valuable ASEAN assets under duress, it is worth stressing that many of the measures currently being taken by individual governments and ASEAN as a whole are merely the acceleration of existing policy trends. For example, AFTA was inaugurated in the mid-1990s as a means of creating a large, unified ASEAN market which would be attractive to TNCs moving away from reliance on export-orientation. The survey confirmed – see Table 3.10 – that a large potential ASEAN market was indeed of future investment interest to TNCs, and this is likely to be the case again as the situation matures. Indeed TNCs in ASEAN countries more affected by the crisis will find it beneficial to export to other less-affected countries in the region. Similarly, the principal provisions of the ASEAN Investment Area are designed to remove the many bottlenecks inhibiting FDI/TNC operations in and across the region. These provisions include:

- Opening up all industries, with some exceptions...for investment to ASEAN investors (including many foreign TNCs based in ASEAN) by 2010 and to all investors by 2020.
- Granting national treatment, with some exceptions...to ASEAN investors by 2010 and to all investors by 2020.
- Promoting freer flow of capital, skilled labour and professionals, and technology among ASEAN countries.
- Providing transparency of investment policies, rules, procedures and administrative processes.
- Providing a more streamlined and simplified investment process.

These provisions need further elaboration and negotiation among the ASEAN member countries, but it is important to recognise that *they are not* the direct result of the Asian financial crisis. In fact these problems were identified by, among others, the Bradford ASEAN Survey (Mirza *et al*., 1997) and had been fed into the process that was to lead to the AIA well before the commencement of the crisis in July 1997. In other words, TNCs were clearly aware of the obstacles and impediments (e.g. poor physical and commercial infrastructure, lack of quality management and skills, intra-ASEAN non-tariff barriers) they were increasingly facing in the ASEAN region and were actively lobbying to have these ameliorated. Clearly, the long-term prospects were deemed bright enough to continue with considerable investment, despite these short term problems (Mirza and Bartels, 1999). In this context, since the impediments to TNC investors in ASEAN are in the process of being reduced, and a policy framework is being put in place to increase ASEAN's attractiveness,[9] arguably the conditions now exist for a renewal of corporate interest in the region. At the moment, most surveys or analyses, however provisional, tend to support this conclusion (Mirza and Wee, 2000; UNCTAD, 1998a; Arthur Andersen, 1998).

CONCLUDING REMARKS

The results of the above discussion imply that ASEAN's best long-run strategy for attracting non-ASEAN FDI by transnational corporations is to facilitate intra-ASEAN FDI. This makes sense at a number of levels. First, ASEAN is a region which is having to compete (and cooperate) with other regions and growth zones, including Mercosur, 'Greater China', and India in the developing world alone: it therefore needs to stress its critical mass as a community of closely cooperating economies as opposed to a club of individual and individualistic nation states. Second, ASEAN – or at least parts of ASEAN – is maturing and (until recently, prior to the Asian financial crisis) represented a growing market to which TNCs were responding, often by taking advantage of the regional division of labour: this is a natural process and needs to be encouraged. Third, as ASEAN matures, so do its home-grown TNCs which, apart from also pursuing a regional division of labour, are potential targets or partners for non-ASEAN TNCs or their subsidiaries in the region. These tendencies explain the success of the ASEAN Free Trade Area (AFTA), although TNCs in the Bradford survey wanted its completion faster; they also require that the ASEAN Investment Area (AIA) provide a broad physical and commercial infrastructure within which intra-ASEAN trade and investment may occur, as well as a range of measures to remove impediments and bottlenecks. These secular trends and national/regional policies, taken together, are encouraging increased inward investment from outside the region (especially from EU TNCs); and there is some, limited evidence of the intra-ASEAN diffusion of FDI by TNCs accelerating as a consequence of the crisis.

However, this should not be regarded as a foregone conclusion. Since July 1997 the current account balance of the five 'crisis countries' (Indonesia, Korea, Malaysia, the Philippines, and Thailand – four are ASEAN members) has shifted from a collective deficit of $5 bn to a surplus of $60 bn (October 1998); in the meantime the average GDP growth rate for these countries has slipped to −8 per cent. In other words, as export orientation (especially by TNC manufacturers in Asia/ASEAN) increases, the size of the local market is contracting. The case for an increased (ASEAN) regional orientation on the part of TNCs is therefore difficult to sustain if companies look to short-term market signals, especially since ASEAN as a whole is unlikely to resume a positive growth path until the year 2000 at the earliest (World Bank, 1999). Having said this, ASEAN is a variegated region: some countries have come through the crisis better; some have reasonably-sized growing markets; and the factor endowments across Southeast Asia are so disparate that a regional division of labour 'positively' encourages intra-ASEAN FDI (by both ASEAN and non-ASEAN TNCs) and the development of integrated production networks. There is hope then, but also a challenge for governments and TNCs alike.

NOTES

1. The survey is hereafter referred to as the 'Bradford ASEAN survey': the data are collated in the 'Bradford ASEAN Foreign Direct Investment Database'. At an early stage it was decided to restrict the survey to *manufacturing* TNCs because they face markedly different contingencies

from, say, international investors in the services (although a few non-manufacturing firms were captured in the survey). In order to make the survey comprehensive it was decided to try and survey as many source countries as possible. In the event, 245 postal questionnaires were returned by parent TNCs in non-ASEAN countries (20 European, 36 US, 140 Japanese, 37 Taiwanese, 13 Australian, and 2 Other).

The 20 page questionnaire consisted of six parts: (i) background information (source country, industry, details of company, organization, views on ASEAN etc.); (ii) investment, motivation and strategy (subsidiary location, reasons for investment, customers, degree of integration and links with local firms); (iii) future investment intentions (parent's commitment, priority markets, plans in ASEAN); (iv) management and commitment (staffing, commitment to specific activities, partners); (v) performance and impact (success factors, technology transfer, business culture transfer, linkages); and (vi) policy matters (obstacles, hindrances, ways of enhancing ASEAN's attractiveness). Many questions are multiple level (e.g. host and general ASEAN perceptions, present and future practices).

Some Characteristics of the Firms analysed in this chapter

Item	Japanese Firms	Taiwanese Firms	Australian Firms	European Firms	US Firms
Number of firms	89*	34	13	19	23*
Mean Employment	7,200	2,237	19,900	46,500	52,300
Median Employment	3,000	1,700 (17,000**)	40,000	16,000	50,000
Three Most Common Industries	Automobiles, Electronics, Metal Products	Electronics, Metal Parts, Rubber & Plastic Products	Automobile Components, Food, Financial Services	Telecomm., Petrochem., Speciality Chemicals	Telecomm., Petrochem., Speciality Chemicals

Notes: * The responses from 89 firms were analysed for this chapter, though more took part in the survey.
  ** One parent company employed 33,800, and was an outlier which distorted the averages. The median of 1700 excludes this company.
Source: The Bradford ASEAN Foreign Direct Investment Database.

The above table presents the main characteristics of the firms analysed in this paper. Two points of significance need to be made. First, there is a big size gap between firms from different countries: for example, the average size of Japanese and Taiwanese firms is only a fraction of the average size of the US firms. The low median also implies a greater proportion of SMEs in the Japanese and Taiwanese samples. The other big difference is in the most common industries of investment:

European and US firms tend towards telecommunications, petro-chemicals and speciality chemicals; Japanese firms are more frequently found in automobiles, electronics and metal products; Taiwanese firms mainly invest in electronics, metal products and rubber and plastic products; and Australian investment is relatively 'idiosyncratic' – the sample TNCs are to be found in automobile components, food, and financial services. These variations (and others) influence the strategy, behaviour, and responses of firms from each of these three industries and will be touched on in the chapter.

2. This trend has recently reversed, with Latin American countries making a comeback in terms of receiving FDI.

3. The figure for 'other', principally Asian, investors in Table 3.3 is probably an overestimate because accurate data are not kept by all Asian sources of FDI; on the other hand not all Asian FDI in ASEAN is officially reported. Of course, some of the 'other' investment is also intra-ASEAN investment.

4. These figures are taken from a number of country papers presented at the Experts Seminar on Promotion of Foreign Direct Investment in the Context of the ASEAN Investment Area, 23–24 May 1996, Bangkok. The seminar was organized by ASEAN and the UNDP, and kindly hosted by the Thai Board of Investment.

5. The People's Republic of China, Taiwan, and Hong Kong.

6. Low-cost labour seems to be less important in the current circumstances, but Taiwanese firms are not included in the table.

7. KPMG corporate finance data, as reported in UNCTAD (1998b: 5).

8. The reason for this desirability can best be understood in terms of one of the fundamental causes of the Asian financial crisis: the role of China in the Asian and world economy. The original report, based on the Bradford ASEAN Survey (Mirza *et al.*, 1997), suggested that, despite ASEAN's fears that China was sucking away FDI, many companies still viewed the ASEAN countries as prime targets. The report suggested that China did not matter so much: 'Why was the China factor not so sig-nificant?' The answer is multi-fold, although there is a need to analyse the results further: (a) China is bigger in terms of population, but the total market size of ASEAN compensated for this because of higher per capita incomes; (b) FDI in Asia, including ASEAN, is increasingly market orientated (be this in consumer goods, industrial goods or infra-structural projects) therefore cost is a less significant factor: local ties and relationships are much more important; (c) there are very cheap investment locations within ASEAN and these are easily accessible: China may increasingly find that its relative lack of infrastructure will impede further export-orientated FDI – also its resources are very far into the hinterland and hard/expensive to access; (d) many TNCs are well established in ASEAN with excellent local conditions and a (formerly) booming regional market: ASEAN is therefore seen as an integral regional economy in its own right; and (e) note also that many

companies are using ASEAN as an integrated base for some global products (e.g. air conditioners): in this sense there will be trade within any TNCs internal network between, say China, ASEAN and India. (With regard to the last point, many TNCs in ASEAN and ASEAN firms themselves are already investing in China and India – this is not a bad thing: opportunities are being realized and integration between the sub-Asian regions is becoming a reality. Chinese companies are also already investing in ASEAN.)' In the light of recent events, however, the above conclusion needs to be qualified because in one fundamental sense China 'does matter'. It matters because, by becoming a massive and cheap producer of a whole range of products, China (along with other Asian countries) has helped to create a situation of global oversupply (deflation) which has affected countries throughout the rest of Asia – and further afield. This was a fundamental factor in the genesis of the Asian financial and economic crisis: the economies and balance of trade of countries such as Thailand suffered because of competition from China, thus creating jitters among foreign investors. A major global (not just Asian or ASEAN) problem, then, is how to encourage further consumption in the world economy in order to 'mop up' this oversupply. Inevitably, much of the additional consumption must come from China, ASEAN, and the rest of the developing world. Ironically, by increasingly becoming a market for TNC products, ASEAN was already moving towards this situation (as indicated by the above quote). Measures are needed to boost consumption (e.g. through fiscal measures, a rise in wages etc.) in ASEAN, China, and the rest of Asia: AFTA and the AIA have a role to play in the process.

9. The report (Mirza *et al.*, 1997) found six factors particularly pertinent for increased ASEAN attractiveness: enhanced specific ASEAN-wide assistance and schemes, regional political and economic stability, ASEAN-wide FDI policies, the completion of AFTA, ASEAN-wide risk reduction schemes especially for SMEs and ASEAN-wide information, and market services especially for SMEs.

BIBLIOGRAPHY

Arthur Andersen/Ministry of Economic Affairs (France) (1997) *International Investment Towards the Year 2001* (Paris: Arthur Andersen).

Arthur Andersen/Ministry of Economic Affairs (France) (1998) *International Investment Towards the Year 2002* (Paris: Arthur Andersen).

ASEAN Secretariat (1998) *Handbook of Investment Agreements in ASEAN* (Jakarta: ASEAN Secretariat).

European Commission/UNCTAD (1996) *Investing in Asia's Dynamism: European Union Direct Investment in Asia* (Luxembourg: Office for Official Publications of the European Communities).

Giroud, A. (1999) 'Manufacturing in The Asia Pacific: East, Southeast Asia and the Pacific'. In Malcolm Warner (ed.), *International Encyclopaedia of Business and Management: Regional Set, Volume 4 – Countries in the Asia Pacific*, pp. 156–67 (International Thomson Business Press).

Japan Institute for Overseas Investment (1997) 'Implications of Regional Trade Liberalisation in ASEAN Countries for Japanese Companies in the Region', *JOI Review*, September.

Mirza, H. (1998) *Global Competitive Strategies in the New World Economy: Multilateralism, Regionalisation and the Transnational Firm* (London: Edward Elgar).

Mirza, H. and Bartels, F. (1999) 'Signals of the Asian Crisis: Obstacles and Managerial Impediments to Foreign Direct Investment in ASEAN'. Paper presented at the AIB Asia Pacific Region Conference, Sydney, July.

Mirza, H. and Kee Hwee Wee (2000) *Transnational Corporate Strategies in the ASEAN Region* (London: Edward Elgar).

Mirza, H., Bartels, F., Hiley, M. and Giroud, A. (1997) *The Promotion of Foreign Direct Investment into and within ASEAN: Towards the Establishment of an ASEAN Investment Area* (Jakarta: ASEAN Secretariat).

UNCTAD (1996) *Sharing Asia's Dynamism: Asian Direct Investment in the European Union* (New York: United Nations).

UNCTAD (1998a) *World Investment Report 1998: Trends and Determinants* (New York: United Nations).

UNCTAD (1998b) *The Financial Crisis in Asia and Foreign Direct Investment: an Assessment* (Geneva: UNCTAD).

World Bank (1999) *Global Economic Prospects and the Developing Countries 1998/99: Beyond Financial Crisis* (Washington DC: World Bank).

4 ASEAN's Outward Direct Investment in Europe

Jim Slater

INTRODUCTION

Prior to the 1997 currency crisis in Southeast Asia, outward direct investment from the region had been conspicuously on the increase. UNCTAD (1997a) reported that, among the leading outward investors, the ratio of ODI to gross fixed capital formation averaged 9.5 per cent for Singapore, and 6.9 per cent for Malaysia, over the period 1991–95. These figures compare with an average of 5.6 per cent for all developed countries, 7.9 per cent for the European Union, and 6.6 per cent for the United States. Some Asian governments were, and are still, actively encouraging ODI, and the scale of existing and potential movements of capital have made a significant impact on global aggregate flows.

Flows from 'developing Asia' to Europe grew from an annual average of US$100 m during 1989–91 to US$5 bn during the early 1990s (UNCTAD, 1997a).[1] Even so, the cumulative ODI total in Europe still only comprised about 4 per cent of Asia's total ODI stock. In contrast, investment from the European Union (EU) accounted for 28 per cent of investment flows to 'developing Asia' over the period 1992–95, dwarfing the 9 per cent share of Japan and the 5 per cent share of the United States (US).[2] Europe's relative lack of involvement as a host country for Asian ODI began to be a matter of concern to European policymakers, and the issue was a significant agenda item at the Asia–Europe Meeting (ASEM) in Bangkok in July 1996.[3]

This chapter is an attempt to assess the significance of ODI in Europe from the five main ASEAN (ASEAN-5) countries.[4] Any such attempt, however, immediately faces the problem that there are no regularly published, consistent statistics of inter-country or inter-region ODI stocks or flows. In this chapter, therefore, we try to pull together data from a variety of sources – home country and host country, on ODI flows and stocks, aggregate statistics and firm-level information – to present as comprehensive a picture as possible given the data deficiencies.[5]

The structure of the chapter is as follows. The first section outlines the policies of the ASEAN-5 governments with regard to outward direct investment, and examines the implications of these policies for the reporting of ODI and the collection of ODI statistics. The following section gathers together data on ODI flows from ASEAN to Western Europe. Two sources are used: home (ASEAN) country data (where available) and data provided by the major host (European) countries. The next section contains a similar analysis, but using data on the stock of ASEAN ODI. Comprehensive data by industrial sector and by host country are provided for Singaporean ODI, but are not available for the other four ASEAN countries. The penultimate section then reports an analysis of a count of overseas affiliates from the ASEAN-4 countries.[6] The final section summarizes the findings, and speculates on the future development of ASEAN ODI in Europe.

GOVERNMENT POLICIES ON OUTWARD INVESTMENT

We first examine government policy towards ODI in each of the ASEAN-5 economies. Not only does such policy affect the flows of ODI directly, but it also has a bearing on what official statistics are collected or, as we will see in the cases of Indonesia and the Philippines, not collected.

Indonesia

Lecraw (1995) has documented Indonesian Government policies with regard to foreign direct investment (FDI). He concluded that both conscious regulation and possibly unintentional intervention have, directly or indirectly, radically influenced the *inflow* of direct investment. Intervention has been extensive and has been substantial at the micro-level, as well as consequential upon attempts to manage macro-aggregates. Fluctuations in liberalization/restriction have clearly increased transnational corporations' (TNCs) perceptions of the political risks associated with investment in Indonesia. Broadly, Lecraw concludes that Indonesian government policies have had adverse effects on inward FDI.

In an earlier paper Lecraw (1993) put the prime focus upon outward FDI. Unlike inward FDI, *outward* FDI from Indonesia has been subject to no direct government interference. Indeed, during the last twenty years, Indonesia has enjoyed a foreign exchange system with outflows totally unrestricted. There has been no reporting requirements and, therefore, no publicly available data from the authorities, so that the ODI figures, such as they are, are appallingly inaccurate. The outward investment figures for Indonesia quoted in the annual *World Investment Reports* were, until 1992, simply the ODI (as recorded by the host countries) having entered the United States and the European Union. Indonesia's outward stock figure in more recent editions of the *World Investment Report* is estimated by the stock held in the United States in 1993, to which has been added subsequent annual investment flows.

Government activity has affected outward investment indirectly, primarily through trade and industry policies. Through the 1980s, these were fundamentally protective in an attempt to realize import-substitution. Protected markets, particularly in manufacturing, led to easy profits with little incentive to invest overseas. However, towards the end of the decade, policy shifted towards export promotion. Devaluations, dismantling of trade barriers, and industry deregulations stimulated export-orientation. In 1988, deregulation of the financial sector increased the capital available for overseas investment. This, combined with the competitive forces unleashed by trade liberalization, stimulated some companies to seek resources abroad to consolidate export markets or to defend domestic markets. There is also some evidence (Lecraw and Todino, 1994) that the consolidation of the EU and NAFTA trade blocs in the early 1990s was anticipated by some Indonesian companies and that this led to some precautionary investment. As with the other ASEAN countries, most ODI has been intra-regional. Of that which has survived the initial surge in the late 1980s to early 1990s, much has been directed into China. Investment into the developed economies appears to have been largely directed to the United States and, in Europe, to the Netherlands.

Although Indonesia's ODI policy seems to have been largely passive, there have been diplomatic initiatives leading to some 35 bilateral investment treaties as at 1 January 1997 (UNCTAD, 1997a), of which 13 have been with countries in Western Europe and eight with countries in Central and Eastern Europe (CEE).

The Philippines

In recent years, the Philippines has been the least active of the ASEAN countries in promoting ODI. As in Indonesia, recording of outward investment is incomplete, reflecting only a partial reporting requirement. Investments in excess of US$6 m per investor per year, using exchange purchased from local banks, need to be registered/ approved by the Central Bank. If the investors are the financial institutions themselves, there is no such requirement. Therefore, whilst domestic figures exist (unlike Indonesia) they substantially under-report outflows. Whilst ODI and foreign exchange has been relatively uncontrolled, the Filipino Government had signed only 16 bilateral treaties (six with Western Europe, two with CEE countries) at 1 January 1997, the lowest number among the ASEAN-5. Political turmoil and lack of restrictions probably stimulated a flight of capital, largely unreported domestically. Sizeable stocks are, however, recorded in developing countries. Among the developed countries, the United States is the largest recipient. Recently, small flows have been recorded into Benelux and France.

Thailand

A process of liberalization of foreign exchange and capital transfers took place between 1991 and 1994. In 1994, the reporting/approval requirements for overseas investments were raised from US$5 m to US$10 m per resident individual per year. The controls have at least meant that, since 1978, Thailand has been collecting ODI data, albeit imperfect for the usual reasons. The Thai Government has been fairly proactive recently: by 1 January 1997, some 20 bilateral treaties had been signed (five in Western Europe, five in CEE countries). Moreover, the Board of Investment has been operating policies to assist Thai firms seeking opportunities abroad. Measures include trade missions and a data focus for potential partners and regulations in host countries. The emphasis has been on ODI into neighbouring countries. Broadly, Thai ODI, though geographically widespread, has concentrated in developing countries. Among the developed countries, the United States has been the leading recipient, followed by Australia. Denmark, Germany, the United Kingdom, and Belgium/Luxembourg bring up the thick end of the tail in Europe.

Malaysia

Singapore apart, Malaysia has had the most outwardly-oriented investment regime among the ASEAN countries. It derives from Prime Minister Mahathir's firm vision of Malaysia's economic development and global political influence. Diplomatic initiatives, generally involving personal visits by Mahathir accompanied by a retinue of senior politicians and representatives of Malaysian business, have encompassed developing countries in South America, Southern Africa, and the 'Muslim Market' in addition to relatively close neighbours. Malaysia had bilateral treaties with 40 countries at 1 January 1997, of which thirteen were with Western European countries and eight with CEE states. A significant number of Memoranda of Understanding have also been signed. That overseas investments have required approval and still do so (for annual investments exceeding US$4 m in 1996) means that ODI statistics have been available on a home country basis in Malaysia for some time. However (as we argue later), there is likely to have been considerable under-reporting of investment flows.

In addition to political encouragement and facilitation, Malaysia has also offered domestic overseas investors a package of financial incentives, including tax write-offs and exemptions. These are designed *inter alia* to favour repatriation of earnings, so as to ensure long-term benefit to the balance of payments. The schemes are based largely on policies enacted and implemented earlier in Singapore. Despite the apparent rivalry between Malaysia and Singapore, which occasionally flares into open rows, there is a commonality of world view at senior political levels. Indeed, in November 1995, a venture capital fund was set up to aid companies from both countries to jointly explore opportunities in third markets.

Whilst much of the investment is destined (at least initially) for Singapore and much of the identifiable remainder rests within the region, the ODI directed to the developed countries manifests a different pattern from that of Indonesia, Thailand, and the Philippines. Japan and Australia were the leading destinations until the mid-1980s, with the United States lagging behind Europe. Since then, Europe still seems to have maintained its attraction to Malaysian firms in flow terms, apart from one or two years. Within Europe, the United Kingdom has proven the favoured target. An increasing number of projects have been undertaken in the CEE countries (Slater, 2000).

Singapore

Singapore's economy is extremely open in terms of both trade and investment. The city-state is by far the largest outward investor of the ASEAN-5 economies, with stocks more than double those of Malaysia. Its entrepôt role complicates the interpretation of statistics, since a high proportion of the outward investment is either reinvestment from external sources or is undertaken by foreign-owned companies. In recent years, Singapore's Department of Statistics has produced a detailed breakdown of investment data which provide a good analytical basis. Firms are required to submit detailed returns and, from 1994, the survey base was expanded to include the financial services sector. The figures show not only the destination of ODI, but also provide breakdowns by activity at home and abroad, and by ownership (Singaporean or overseas). Firms are required to report stocks, and flow data are estimated by the differences in stocks.

The Singaporean authorities have adopted an active, dirigiste, strategic attitude to outward direct investment. The aim has been primarily to secure Singapore's role as a gateway to the region – 'bringing the world to the region and the region to the world' – whilst simultaneously enhancing the competitiveness of Singaporean companies through developing capabilities via exposure to international operations. Policies have been designed with the awareness that ODI should complement the domestic economy rather than encouraging 'hollowing-out'. There have been a variety of schemes administered by the Economic Development Board (EDB) under a series of programme headings. The Regionalization programme was launched in 1992 and its remit was broad enough to appeal both to companies already established in Singapore (whether locally or foreign-owned) and to inward investors with the interest in joining the 3000 or so TNCs trading in Singapore. The financial packages which have since been developed include:

- The Overseas Investment Incentive: provides tax write-offs on Singaporean income against overseas capital loss.
- The Overseas Enterprise Incentive: introduced in 1996 to encourage home-grown TNCs. This exempts foreign-sourced income from tax for five years. Initially this was aimed at ventures with 100 per cent Singaporean shareholding, reduced to 50 per cent in 1997, and waiver possibilities on a case-by-case basis.
- The Regionalization Training Scheme: provides companies with permits and grants for training overseas workers in Singapore.
- The Regional Venture Funds Scheme: provides capital to local companies venturing overseas.
- Investments Study Grants: provides funds for feasibility studies.
- The Double Deduction for Overseas Development Expenditure Scheme: provides awards for companies to explore and study regional business opportunities.
- The Cluster Development Fund: originally established with a budget of S$1 bn, the Fund finances co-investment between the EDB's investment arm and TNCs or 'promising local enterprises' in key projects in both Singapore and the region. The Fund was expanded to S$2 bn in 1997, and is targeted to particular strategic sectors.

European companies have been involved in several of these schemes through various joint operations, ranging from specific projects such as the Daimler Benz–Sembawang Group – Autostar EDB light vehicle venture in Vietnam, to participation in the seven flagship industrial park projects in Indonesia, Vietnam, China, and India. The largest of these is the Suzhou project in China, with investment commitments totalling about US$3 bn. The focus of the policies is, however, regional and, whilst they are spread geographically from Oman and India in the West to China in the East, they are all within Asia.

Somewhat surprisingly, Singapore (at January 1997) had concluded only 15 bilateral investment treaties, six in Western Europe, seven within Asia, and two with CEE countries. However, this almost certainly indicates focused strategic intent rather than window-dressing. Although Europe has not benefited directly from the ODI incentive programmes Singapore has recently taken, steps have been taken to bring European and Singaporean firms together. ASEM Connect derives from the private-sector component of the ASEM process and is an Internet based network of company databases providing consultancy, matchmaking, and business information services. The Singapore Government has also established bilateral business councils with several European countries (German–Singapore Business Forum, France–Singapore Business Council, and the Singapore–British Business Council) though, again, the aim is to encourage European TNCs through the Singapore gateway.

FLOWS OF OUTWARD DIRECT INVESTMENT FROM THE ASEAN COUNTRIES

Table 4.1 provides data on total flows of ODI from the ASEAN-5 countries over the period 1985–97. The data are collated by UNCTAD, and show a clear upward trend for all five countries. Stock data are also given. Unfortunately, however, UNCTAD do not regularly publish a breakdown of these ODI flows by destination, so it is not possible to identify how much of this investment is finding its way to Europe. Alternative sources of data must therefore be sought.

Disaggregated data are published on ODI to Western Europe by the authorities in Thailand, Malaysia, and Singapore – see series A in Table 4.2. But, as noted in the previous section, no such data are either collated or published in Indonesia or the Philippines. Another possibility is thus to use figures collated by the recipient countries, as recorded by the OECD (1997)[7] – see series B in Table 4.2. Negative signs imply disinvestment, or reverse investment. FDI data normally include three components when collated for balance-of-payments purposes: equity capital, reinvested earnings, and intra-company loans. A negative FDI flow implies that at least one of these components is negative and outweighs the others. However, it is worth noting that repatriation of earnings or the repayment of loans do not necessarily imply a reversal of the investment commitment.

One obvious point to note about the data in Table 4.2 is the disparity between the two series, though there is some evidence (particularly from series A) that FDI in Europe is on the increase. The discrepancies between the two series may reflect, in part, differences

Table 4.1   Outward Direct Investment from Five ASEAN Countries, 1985–97[1]
(US$m)

Year	Malaysia		Thailand		Singapore			Indonesia		Philippines	
	Flow	Stock	Flow	Stock	Flow	Stock[2]	Stock[3]	Flow	Stock	Flow	Stock
1985	210	749	1	14	238	6254	1026	33	49	−3	171
1986	249	770	1	16	181	6435	1193	−11	37	−5	157
1987	209	1023	170	189	206	6641	1406	−5	35	9	156
1988	206	1469	24	212	118	6759	1488	26	41	13	154
1989	282	1751	50	258	882	7641	2712	17	39	6	160
1990	532	2283	140	398	2034	9675	7513	−11	25	−5	155
1991	389	2672	167	565	526	—	8787	13	42	−26	129
1992	514	3186	147	712	1317	—	10891	52	101	5	134
1993	1325	4511	233	945	2021	—	13726	356	83	374	508
1994	1817	6328	493	1438	3746	—	18489	609	98	302	810
1995	2575	8903	886	2324	3988	32695	24792	603	1295	98	909
1996	3700	12603	931	3255	4805	37500	n.a.	512	1809	182	1091
1997	3100	15703	500	3755	5900	43400	n.a.	2400	4207	136	1227

Notes: (1) The 1997 figures are estimates.
(2) This incomplete stock series is compiled by UNCTAD. The figures for 1995–97 are not comparable with the earlier data.
(3) This stock series is reported by the Department of Statistics, Republic of Singapore and relates to 'direct equity investment'.

Sources: 1985–88: IMF, *International Financial Statistics*.
1989–97: UNCTAD (1995, 1996a, 1997a, 1998).

Table 4.2　Outward Direct Investment Flows from Five ASEAN Countries to Western Europe, 1985–96[1] (US$m)

Year	Malaysia		Thailand		Singapore		Indonesia	Philippines
	A	B	A	B	A	B	B	B
1985	22.84	1.02	0.04	-5.10	35	—	0.34	—
1986	20.65	0.92	0.32	-7.43	89	—	1.39	0.14
1987	28.26	0.80	0.00	0.63	-21	6.01	0.93	0.17
1988	26.20	-277.50	0.00	-0.15	-43	11.34	8.51	7.65
1989	43.49	2.81	0.03	-0.33	118	38.89	15.02	0.50
1990	50.39	-3.76	12.48	0.91	346	31.70	1.95	0.09
1991	46.58	3.64	15.88	2.48	205	24.02	4.32	0.79
1992	55.04	-2.58	11.07	2.05	100	132.36	16.89	14.30
1993	215.23	104.16	13.04	14.39	121	117.36	-13.97	1.07
1994	306.40	-132.97	23.38	51.36	472	-48.39	4.23	1.57
1995	240.97	43.62	152.94	14.95	1271[2]	-184.30	-1.26	1.83
1996	303.23	n.a.	13.05	n.a.	n.a.	n.a.	n.a.	n.a.
Averages								
1985–90	31.98	-45.95	2.15	1.43	94	14.66	4.52	1.71
1991–96	194.58	2.02	38.23	14.36	423[3]	12.13	2.03	3.28

Notes:　(1)　Series A are based on data provided by the home (ASEAN) country; Series B are based on data provided by the major host (European) countries.
　　　　(2)　The large increase over the previous year reflects an expanded survey base.
　　　　(3)　The average figure is for the years 1991–95 only.

Sources:　Series A: Bank Negara, Malaysia; Bank of Thailand; Department of Statistics, Republic of Singapore.
　　　　Series B: OECD (1997).

in the criteria for data collection. Thus, the series *A* data for Thailand are based upon equity flows plus repatriation of capital, and do not include long-term loans. The Malaysian data are for approvals, rather than actual implemented investment. The Singapore data conform to IMF guidelines, but are derived from surveys. In contrast, the data in Series *B*: (a) refer to actual flows (including long-term loans) rather than approvals; and (b) much is directed to the United Kingdom (see below) where the threshold for reporting inward FDI is an equity holding of 20 per cent, compared to the international norm of 10 per cent.

Given the year-on-year volatility of the FDI flows, a more useful indicator of shifts in the attitudes of Asian investors is to express the investment into Europe as a proportion of the ASEAN countries' total ODI[8] – see Series *C* in Table 4.3. But here the picture is complicated by the role of Singapore as a regional financial centre. On the one hand, some recorded ODI in Singapore represents primarily portfolio investment in holding companies, and thus overestimates the true flow of ODI. On the other hand, such holding companies may act as vehicles for ODI in other countries, thus overestimating the flow of ODI to Singapore and underestimating the flow to the other countries. These distortions are likely to be particularly severe for neighbouring Malaysia and Indonesia. Thus we also report ASEAN investment in Europe as a proportion of total ODI excluding investment in Singapore – see Series *D* in Table 4.3.

Notwithstanding these adjustments, there is still considerable volatility in the data and trends are difficult to pick out. Neither Indonesia nor the Philippines appears to have much investment in Europe. But it appears as though Malaysian, Thai, and Singaporean businesses showed increasing interest in Europe over the decade from 1985 to 1995, and the ASEAN investment boom in the early to mid-1990s neither favoured Western Europe nor passed it by. The reasons for the boom are not the subject of this chapter but, with hindsight, a flight of capital in anticipation of, and reinforcing the conditions for, a currency crisis is a possibility. As Thurow (1998) points out in his analysis of the crisis, those best placed to anticipate the currency crisis were those with local knowledge. How much of the 1994–95 surge in ODI was translated into genuine overseas business activity is an interesting question. An examination of the detailed Thai statistics, for example, shows a sudden increase in ODI in

Table 4.3   Outward Direct Investment Flows from Five ASEAN Countries to Western Europe, as Shares of Total ODI Flows, 1985–95 (% of total outward direct investment)

Year	Malaysia		Thailand		Indonesia		Philippines		Singapore
	C	D	C	D	C	D	C	D	C
1985	11.4	17.5	1.8	1.8	0.6	5.3	0.0	0.0	20.5
1986	11.7	29.3	28.1	28.1	1.8	8.5	0.5	9.8	53.0
1987	13.4	19.5	0.0	0.0	1.0	11.6	0.5	7.2	−164.4
1988	13.6	15.7	0.0	0.0	7.9	14.3	17.1	25.7	n.a.
1989	18.7	23.7	0.1	0.1	10.1	22.4	0.4	1.5	n.a.
1990	9.8	15.0	9.0	9.1	0.4	0.5	0.0	0.0	6.3
1991	12.4	18.8	9.3	11.1	0.8	0.9	1.0	8.2	19.3
1992	10.9	13.7	7.6	8.7	4.4	4.9	11.7	47.1	3.2
1993	17.2	21.8	4.6	5.0	−7.3	−16.2	1.1	n.a.	1.9
1994	14.7	17.8	5.6	6.2	n.a.	7.7	n.a.	2.4	10.7
1995	9.2	13.5	19.4	19.5	n.a.	n.a.	n.a.	4.9	23.8

Notes: (1) Series *C* shows ODI in Europe as a percentage of total ODI by the country indicated; Series *D* shows ODI in Europe as a percentage of total ODI, less any ODI in Singapore, by the country indicated.
　　　　 (2) The figures for Malaysia, Thailand, and Singapore are based on home country data; the figures for Indonesia and the Philippines are based on host country data.
　　　　 (3) Negative figures indicate disinvestment.
Sources: Bank Negara, Malaysia; Bank of Thailand; Department of Statistics, Republic of Singapore; OECD (1997).

havens such as the Cayman Islands and the British Virgin Islands. A more prosaic explanation might emphasize the tax holidays offered in Singapore and Malaysia, which reinforced the increasingly outward-looking orientation of their companies.

STOCKS OF OUTWARD DIRECT INVESTMENT FROM THE ASEAN COUNTRIES

An alternative perspective on the ASEAN commitment to Europe may be gathered from a scrutiny of the ODI stock data. As before, we report the ASEAN ODI stock in Europe as a percentage of the total

Table 4.4   ASEAN Outward Direct Investment Stocks in Western
Europe, as Shares of Total ODI Stocks, 1980–95
(% of total outward direct investment stock)

Year	Malaysia		Thailand		Indonesia		Philippines		Singapore
	E	F	E	F	E	F	E	F	E
1980	0.0	0.0	8.0	21.0	1.0	6.0	0.0	0.0	3.1
1992	0.0	2.0	2.0	11.0	20.0	24.0	4.0	25.0	8.3
1995	1.0	10.0	6.0	32.0		57.0	4.0	24.0	11.0

Notes: Series *E* shows the stock of ODI in Europe as a percentage of total ODI stock of the country indicated; Series *F* shows the stock of ODI in Europe as a percentage of the total ODI stock, less any ODI in Singapore, by the country indicated.

Sources: Bank Negara, Malaysia; Bank of Thailand; Department of Statistics, Republic of Singapore.

ODI stock – see Series *E* in Table 4.4 – and the ASEAN ODI stock in Europe as a percentage of the total ODI stock excluding investment in Singapore – see Series *F* in Table 4.4. The ODI stock of each country has increased substantially over the 15-year period from 1980 to 1995.[9] All the countries (except for Thailand) recorded an increased share of the ODI stock in Europe between 1980 and 1992 and, by 1995, the Thai figure had risen again. Europe therefore seems to be have been an increasingly favoured destination for ASEAN ODI. These trends are all the more dramatic when the shares excluding ODI in Singapore are considered. Europe accounted for 57 per cent of Indonesia's remaining ODI stock in 1995, 32 per cent of Thailand's, 24 per cent of the Philippines', and 10 per cent of Malaysia's. Except for the Philippines, much of the increase in share took place between 1992 and 1995.

The one ASEAN country that does provide detailed statistics on its stock of ODI is Singapore. Table 4.5 shows a breakdown of the ODI stock at the end of 1995, according to sector and host country. Europe accounts for over $5 bn of Singaporean ODI, over twice as much as the United States. Much of the ODI in Europe is located in the United Kingdom and, in a country ranking of favoured locations, the United Kingdom comes fourth behind Malaysia, Hong Kong, and Indonesia. Both Europe and the United States have accounted

Table 4.5 Singapore's Outward Direct Investment in Selected Host Countries, by Sector at end 1995 (S$m)

Country	Manufacturing	Construction	Commerce	Transport	Financial	Real Estate	Business Services	Other	Total
Europe	208	—	108	194	4636	40	312	52	5551
Netherlands	4	—	8	—	1009	—	—	—	1021
United Kingdom	61	—	49	13	2832	18	275	49	3297
United States	122	—	207	20	1914	150	221	2	2635
Australia	221	11	195	5	448	488	43	36	1448
New Zealand	12	—	40	5	1891	127	18	25	2118
Asia	10637	478	3921	568	7914	2434	679	472	27101
Brunei	4	33	15	—	3	—	5	31	92
Indonesia	3003	94	147	96	243	346	36	66	4031
Malaysia	4045	47	1513	131	3182	476	117	205	9716
Philippines	162	86	29	6	154	19	138	30	625
Thailand	629	33	367	39	65	74	26	20	1253
Vietnam	181	2	5	4	125	30	17	8	371
ASEAN	8025	295	2076	276	3772	944	340	361	16088
Hong Kong	701	136	732	191	3399	928	168	13	6268
Taiwan	170	24	307	13	30	12	17	—	573
China	1512	21	495	65	223	520	49	83	2968
Japan	52	—	135	5	211	—	63	—	466
TOTAL	11397	505	4528	1353	23172	3380	1317	588	46240

Source: Department of Statistics, Republic of Singapore, *Singapore's Investment Abroad*.

for roughly stable shares of total Singaporean ODI over the period from 1985 to 1995. Malaysia's share of the total has, however, dropped over the same period, as more ODI has been directed towards China and Indonesia. Over $4 bn has been invested in two major tax havens – the Netherlands Antilles and the Cayman Islands – particularly since 1995, though this ODI stock is not shown in the table.

Less than 4 per cent of Singaporean ODI in Europe is in manufacturing industry, whereas almost 84 per cent is in the financial sector. In contrast, about 50 per cent of Singaporean ODI in Asia is in manufacturing, ranging from 11 per cent in Hong Kong, to just over 50 per cent in Thailand and China, to 75 per cent in Indonesia. Australia (ranked ninth in terms of total ODI) reveals a relatively even sectoral spread: 31 per cent in financial services, 34 per cent in real estate, 15 per cent in manufacturing, and 14 per cent in commerce. The ODI in the tax havens is of course in the financial sector, and is larger than the UK and Indonesian totals. Overall therefore, the greatest part of Singapore's ODI is devoted to the financial sector.

AN ANALYSIS OF OUTWARD DIRECT INVESTMENT BY ASEAN COMPANIES

Thus far we have attempted to assess trends in ASEAN ODI using data published for official statistical purposes. Such official data provide some disaggregation by sector or by host country but, with the exception of Singapore, do not provide a breakdown by sector and host country. We have therefore resorted to collating information on individual firms from the ASEAN-4 economies, using Dun & Bradstreet (1997, 1998).

The information in the Dun & Bradstreet survey is largely qualitative, and tends to concentrate on larger companies. Many of the parent companies are registered as holding companies (Indonesia 41 per cent; Malaysia 34 per cent; the Philippines 11 per cent; Thailand 8 per cent), but we have effected a sectoral classification of the activities of their overseas subsidiaries. Where possible, we have distinguished between direct and indirect control, the latter reflecting ultimate ASEAN parentage but with control exercised through intermediaries.

Indonesia

Table 4.6 presents a sectoral analysis of the overseas affiliates of Indonesian parent companies in the Dun & Bradstreet survey. It seems that no Indonesian ODI is (primary) resource-seeking, but that most activity is in financial services and distribution. There has also been some activity in manufacturing, and in transport and communications. Hong Kong and Singapore are the most frequent ODI destinations, with other affiliates spread thinly in a number of different locations. The favoured European destinations are Germany (1 affiliate) and the Netherlands (3), but all four affiliates are confined to the financial sector. Thus a high proportion of Indonesia's ODI (as reported in Table 4.4) seems, on the evidence here, to be in financial services and includes a major tranche by the banking sector, including the Bank of Indonesia (the Central Bank). Most of the ODI from the manufacturing sector (located in Hong Kong, India, Singapore, the Cayman Islands, and the United States) appears to be trade-supporting investment. Other sources indicate manufacturing investment in China (not recorded by Dun & Bradstreet) of around US$1 bn.

In conclusion, Indonesia's outward investment seems mainly to service capital and trade flows, with little potential for domestic producers to reap the benefits from internationalization. However, the recorded manufacturing investment does suggest a preference for high value-added items such as scientific equipment, and for goods such as petroleum products, chemicals, glass products, and textiles. This suggests technology acquisition and/or the securing of strategic sources as possible motives for Indonesian ODI. Little of this investment seems to have taken place in Europe.

Malaysia

Table 4.7 provides a summary of Malaysian affiliates in the Dun & Bradstreet survey. Clearly Malaysia, with 134 companies, is much more active in ODI than Indonesia. Popular industrial sectors in the European Union are construction, wholesaling, and manufacturing. Although 55 companies in the primary sector are registered in the United Kingdom, their operations are mainly in Malaysian palm oil and rubber, and derived products.

Table 4.6 Outward Direct Investment by Indonesian Companies

Sector	Number of Companies	Number of Overseas Associate Companies	Number of Overseas Subsidiary Companies	Number of Subsidiaries in Developed Countries
Agriculture, Forestry & Fishing	0			
Mining	1	0	3 (havens^2)	0
Construction	0			
Manufacturing	8	3	$12 + 1$ (haven^2)	2 (USA)
Transportation & Communications	3	1	7	2 (Japan, Australia)
Wholesale	5	3	4	1 (Netherlands)
Retail	1	1	0	
Finance, Insurance & Real Estate[1]	15	5	$11+1$ (haven^2)	2 (Netherlands, Germany)
Services	0			

Notes: (1) Includes Holding Companies whose activities are unknown.
(2) 'Haven' indicates an overseas subsidiary located in a tax haven.
Source: Tirado-Angel (1998), from Dun & Bradstreet (1997, 1998).

Table 4.7 Outward Direct Investment by Malaysian Companies

Sector	Number of Companies	Number of Overseas Associate Companies	Number of Overseas Subsidiary Companies	Number of Subsidiaries in Developed Countries
Agriculture, Forestry & Fishing	16	18	113 (11 havens)	46 (37 in UK, 4 in Australia, 4 in USA, 1 in New Zealand)
Mining	6	12	33	1 in UK, Australia & Japan
Construction	18	65	236 (10 havens)	80[3]
Manufacturing[2]	53	101	548	See Table 4.8
Transportation & Communications	8	19	49	20[4]
Wholesale	25	56	224 (4 havens)	72[5]
Retail	3	4	50	36 (35 in Australia & 1 in USA)
Finance, Insurance & Real Estate[1]	37	74	350 (12 havens)	91[6]
Services	30	66	271 (33 havens)	64[7]
Total companies in sample	134	133	488	171

Notes: (1) Includes Holding Companies whose activities are unknown.
(2) 'Haven' indicates an overseas subsidiary located in a tax haven.
(3) 20 in USA, 24 in UK, 14 in Australia, 6 in Canada, 3 in France and Italy, 4 in the Netherlands, 1 in Greece, Austria, Germany, Spain, Japan, South Africa, and Belgium.
(4) 6 in the UK, 5 in Australia, 2 in Jersey, Germany, and Monaco, 4 in the Netherlands, and 3 in the USA.
(5) 44 in Australia, 11 in UK, 12 in USA, 1 in New Zealand, 2 in Japan, 2 in Netherlands, 1 in Canada.
(6) 23 in USA, 36 in UK, 6 in Netherlands, 5 in Germany, 14 in Australia, 3 in Japan, 1 in Italy, Switzerland, Portugal, Canada, and Belgium.
(7) 20 in Australia, 25 in UK, 2 in USA, 2 in the Netherlands, 3 in Canada, 3 in New Zealand, 1 in Monaco and Germany.

Source: Tirado-Angel (1998), from Dun & Bradstreet (1997, 1998).

The investment structure of many Malaysian companies is complex. Whilst the majority of parent companies invested directly overseas, about one-third did so via intermediary companies. Indeed there are instances of up to five intermediaries between some parent companies and their identified offspring. The complex holding structures of Malaysian corporations, and the frequent use of Singapore, Hong Kong, and offshore tax havens as locations for holding companies, invite a detailed analysis well beyond the space limitations of this chapter. Here we shall just focus upon the manufacturing sector – see Table 4.8. European destinations are underlined.

The favoured locations within the European Union appear to be the Netherlands, Germany, and the United Kingdom. However, all is not as it seems. Relatively few of the affiliates are held by groups based around a core manufacturing business. Furthermore, much of the activity of manufacturers is to provide trade support, through marketing and distribution. For example, Kimberly Clark supplies paper products to Europe through subsidiaries in five European countries. Hume Industries (via intermediaries in Malaysia and the Netherlands) has established subsidiaries in nine European countries for the distribution of air filters. There is little evidence of Malaysian manufacturing companies finding opportunities for, or deriving benefit from, ODI in Europe. The large recorded investment (see Table 4.2) seems mostly for trade-support and profit-taking, with little in higher value-added or technology-intensive manufacturing. There are, however, some small holdings in specialist (oil exploration and vehicle electronics) engineering.

The Philippines

As with Malaysian companies, Filipino organizational structures are complex, involving cross-holdings and the frequent use of intermediaries based either in the Philippines, Hong Kong, or tax havens. Table 4.9 provides a sectoral analysis of the overseas affiliates of Filipino parent companies in the Dun & Bradstreet survey.

In the primary sector, there is evidence of logging in Indonesia, via Hong Kong, possibly as a response to Filipino Government bans on the logging of old growth. Four of the nine manufacturing companies represented are in food products, whilst two are in construction and engineering materials. Investment into developed countries by the

Table 4.8   Outward Direct Investment by Malaysian Manufacturing Companies

Code	Industry Description	Countries in Order of Importance
311–2	Food Manufacturing	Singapore, Thailand, Hong Kong, *UK*, USA, Solomon Islands, Australia, Vietnam, Indonesia, Papua New Guinea, Brunei, China, *Netherlands*, Philippines, Canada, Japan, Tanzania, Egypt, Mexico, India, Liberia
313	Beverage Manufacturing	Brunei, Hong Kong, Thailand, Singapore
314	Tobacco Manufacturing	Brunei, Singapore
321	Manufacture of Textiles	Singapore, Hong Kong, Australia
322	Manufacture of Wearing Apparel Except Footwear	Singapore, Hong Kong, Sri Lanka, Australia
323	Manufacture of Leather and Products of Leather, Leather Substitutes and Fur, except Footwear and Wearing Apparel	Singapore
331	Manufacture of Wood and Cork Products except Furniture	Singapore, USA, Papua New Guinea, Australia
341	Manufacture of Paper and Paper Products	Singapore, *UK*, Australia, Hong Kong, USA, Bangladesh, China
351	Manufacture of Industrial Chemicals	Singapore
352	Manufacture of Other Chemical Products	Australia, Hong Kong, Singapore, USA, Brunei, Vietnam
354	Manufacture of Miscellaneous Products of Petroleum and Coal	Singapore, South Africa, Hong Kong, Australia, USA, Germany, *UK*, India, Indonesia, Vietnam

(continued)

Table 4.8 *(continued)*

Code	Industry Description	Countries in Order of Importance
355	Manufacture of Rubber Products	Singapore, *UK*, Thailand, Hong Kong, USA, China, Australia, Papua New Guinea, Indonesia, Philippines, *Netherlands*, Vietnam, Canada, Japan, Liberia, Tanzania, Egypt, Solomon Islands
356	Manufacture of Plastic Products	Australia
362	Manufacture of Glass and Glass Products	Singapore
363–9	Manufacture of Non-Metallic Mineral Products	Singapore, Australia, South Africa, New Zealand, *UK*, China, Hong Kong, Philippines, Vietnam
371	Iron and Steel Basic Industries	Singapore, Hong Kong, Japan
372	Non-ferrous Basic Industries	*UK*, Zambia, Nigeria, Kenya, New Zealand, Tanzania, Brunei, Singapore
381	Manufacture of Fabricated Metal Products, except Machinery and Equipment	Singapore, Hong Kong, Australia, *UK*, New Zealand, South Africa, Brunei, Papua New Guinea
382	Manufacture of Machinery except Electrical	Singapore, Philippines
383	Manufacture of Electrical Machinery, Apparatus, Appliances and Supplies	Singapore, USA, Papua New Guinea, China, Japan, Tanzania, Bangladesh, Brunei, Cambodia, Solomon Islands, Hong Kong
384	Manufacture of Transport Equipment	Singapore, Hong Kong, Australia, Brunei, New Zealand, China, Papua New Guinea
385	Manufacture of Professional, Scientific, Measuring and Controlling Equipment	Singapore, Japan, Brunei, China, Philippines, Solomon Islands, Papua New Guinea
390	Other Manufacturing Industries	Hong Kong, Australia, USA, Brunei, Singapore, Vietnam

Source: Tirado-Angel (1998), from Dun & Bradstreet (1997, 1998).

Table 4.9 Outward Direct Investment by Filipino Companies

Sector	Number of Companies	Number of Overseas Associate Companies	Number of Overseas Subsidiary Companies	Number of Subsidiaries In Developed Countries
Agriculture, Forestry & Fishing	2	1	4	0
Mining	1	0	4	3 (Japan, Canada, USA)
Construction	2	0	2	1 (USA)
Manufacturing	9	2	28	7 (5 in USA, 1 in Japan and Canada)
Transportation & Communications	3	0	8	0
Wholesale	2	1	1	0
Retail	0			
Finance, Insurance & Real Estate[1]	13	3	36 (3 havens[2])	11 (6 in USA, 4 in Italy, 1 in UK)
Services	2	0	5	4 (2 in USA, 1 in Japan and Canada)
Total companies in sample	27	5	64	18

Notes: (1) Includes Holding Companies whose activities are unknown.
(2) 'Haven' indicates an overseas subsidiary located in a tax haven.

Source: Tirado-Angel (1998), from Dun & Bradstreet (1997, 1998).

manufacturing companies seems wholly trade-related. Only financial services investments have been identified in Europe, with four bank branches established in Italy and one insurance company, probably engaged in portfolio investment, located in the United Kingdom (via the Bahamas). ODI from the Philippines in the European Union is thus minimal.

Thailand

The Dun & Bradstreet sample comprises 25 companies, of which eight are involved in manufacturing, primarily in food products, beverages, fabricated structural metal products, clothing, and toiletries – see Table 4.10. Only one of these is linked with Europe, and this company with holdings in German brewing. One construction-based company has subsidiaries in Europe (six in Denmark, three in Germany, three in the United Kingdom), but could be regarded as European in origin. Three wholesalers have associates or subsidiaries in the United Kingdom, Spain, Italy, and Germany. Otherwise, Thailand's ODI is in financial services with the major banks well represented. In summary, Thai investment in Europe is primarily trade-supporting, property and portfolio holding with little, though not entirely absent, concern for high value-added or technology-intensive areas.

CONCLUDING REMARKS

In this chapter, we have attempted to examine the scale of ASEAN ODI in Europe. Despite the increasing flows of ODI into Europe and, in some cases, increasing shares of total ASEAN ODI, there is little investment outside financial services. Recent well-publicized cases of ASEAN–EU involvement (such as Proton/Lotus, Virgin/MAS, Laura Ashley/UMI, Genting/Lonrho Lion/Avon – all Malaysian) seem exceptional and, anyway, the rationale for some of these liaisons is not immediately obvious. Singapore's large investments do include significant non-financial services investments, primarily in transportation and telecommunications (e.g. an investment of over S$1 bn by Singapore Telecom in Belgium). Even so, they constitute a small proportion of the total. There is also little evidence of

Table 4.10   Outward Direct Investment by Thai Companies

Sector	Number of Companies	Number of Overseas Associate Companies	Number of Overseas Subsidiary Companies	Number of Subsidiaries in Developed Countries
Agriculture, Forestry & Fishing	0			
Mining	0			
Construction	1	2	18 (3 havens[2])	14[4]
Manufacturing	8	4	16[3]	8[5]
Transportation & Communications	1	0	4	2 (USA)
Wholesale	8	5	21 (1 haven[2])	10[6]
Retail	1	0	1	0
Finance, Insurance & Real Estate[1]	9	8	10 (1 haven[2])	2 (USA)
Services	2	3	2	2 (New Zealand)
Total companies in sample	25	16	62	32

Notes: (1) Includes Holding Companies whose activities are unknown.
(2) 'Haven' indicates an overseas subsidiary located in a tax haven.
(3) 6 in Denmark, 3 in Germany, 3 in the UK, 1 in Norway, 1 in Spain.
(4) Includes 2 companies in Hong Kong and 1 in Singapore where investment is for trade purposes only.
(5) 2 in New Zealand, 4 in the USA and 2 in Germany.
(6) 3 in the USA, 2 in New Zealand, 2 in Spain, 1 in each of UK, Germany and Italy.

Source: Tirado-Angel (1998), from Dun & Bradstreet (1997, 1998).

significant export-substituting manufacturing activity in the European Union.

This situation is, of course, in complete contrast to that of Japan and the Asian NIEs, especially Korea and Taiwan. Partly of course, this is because the ASEAN countries are less far along the Investment Development Path.[10] But it possibly also reflects the style, organization and ethos of many of the larger ASEAN companies. Typically conglomerate, they are frequently extremely diverse associations of many small companies with core earnings in primary commodities or derivatives, unlike the Korean and Japanese conglomerates which are based upon a smaller number of much larger manufacturing bases. Non-financial ODI of the ASEAN countries is primarily directed towards the developing countries. In the case of Malaysia, this has been channelled with direct political support. Consider, for example, the ambitious Malaysian projects in Albania (estimated US$40 m).

It seems fairly clear, at the time of writing, that ASEAN companies are not yet ready for or taking seriously the challenge of investment in Western Europe, preferring low-cost primary or manufacturing investments in developing countries, either 'flying-geese' style projects in the Asian region or in more remote destinations to consolidate potential political alliances. Investment in Europe seems primarily 'profit-seeking' with the same motivation as portfolio investment, offering little in the way of upgrading capabilities in the ASEAN parent companies where the need lies.

NOTES

1. The 'developing Asia' category relates to South, East, and Southeast Asia, and includes India and the Newly Industrializing Economies (NIEs) as well as China and the ASEAN countries.
2. Although the EU investment accounts for a substantial proportion of the inflow of investment to 'developing Asia', it only constitutes a small percentage of total EU ODI. About 3 per cent of the EU stock of ODI was held in Asia at the end of 1993, and only an average of 2 per cent of annual EU ODI flows over the previous ten years had been destined for Asia (UNCTAD, 1996b).
3. The issue featured rather less at the second ASEM meeting in London in April 1998.
4. The ASEAN-5 countries are Singapore, Malaysia, Indonesia, the Philippines, and Thailand.

5.   There are also the well-known conceptual problems of interpretation of foreign direct investment (FDI) data, and practical problems such as different reporting criteria in different countries, varying degrees of enforcement and evasion, and collection difficulties.
6.   The ASEAN-5 countries, excluding Singapore.
7.   These figures are available in domestic currencies for all OECD member countries, and have been converted to US dollars using the exchange rates listed in Annex III of OECD (1997). At the time of writing, the most recent data were for 1995.
8.   The shares for Malaysia, Thailand, and Singapore are based upon home country declarations; whilst those for Indonesia and the Philippines are derived from host country data. Comparisons between the two groups of countries should therefore be made with caution.
9.   The following data relate to the total ODI stock of the ASEAN-5 economies, and have been used in the compilation of Table 4.4:

	1980	*1992*	*1995*
Malaysia	1800	11200	13500
Thailand	80	1240	2060
Indonesia	40	220	
Philippines	280	600	720
Singapore	780	10890	24810

10.   See Dunning and Narula (1996).

BIBLIOGRAPHY

Department of Statistics, Singapore (various dates) *Singapore's Investment Abroad* (various editions).

Dun & Bradstreet (1997) *Who Owns Whom?* (London: Dun & Bradstreet).

Dun & Bradstreet (1998) *Who Owns Whom?* (London: Dun & Bradstreet).

Dunning, J.H. and Narula, R. (1996) 'The Investment Development Path Revisited: Some Emerging Issues'. In J.H. Dunning and R. Narula (eds), *Foreign Direct Investment and Governments: Catalysts for Economic Restructuring*, pp. 1–41. (London: Routledge).

Lecraw, D.J. (1993) 'Outward FDI by Firms in Indonesia: Motivations and Effects', *Journal of International Business Studies*, 24, 3, 589–600.

Lecraw, D.J. (1995) 'Indonesia: the Critical Role of Government'. In J.H. Dunning and R. Narula (eds), *Foreign Direct Investment and Governments: Catalysts for Economic Restructuring*, pp. 316–47 (London: Routledge).

Lecraw, D.J. and Todino, H. (1994) 'The Impact of Europe 1992 on Firms in the ASEAN Region'. Paper presented at the meeting of the Academy of International Business, Boston, MA.

Organization for Economic Co-operation and Development (1996) *Detailed Benchmark Definition of Foreign Direct Investment*. Third Revision (Paris: OECD).

Organization for Economic Co-operation and Development (1997) *OECD International Direct Investment Statistics Yearbook* (Paris: OECD).

Pang Eng Fong (1995) 'Staying Global and Going Regional: Singapore's Inward and Outward Direct Investment'. In Nomura Research Institute and Institute of Southeast Asian Studies, *The New Wave of Foreign Direct Investment In Asia*, pp. 111–29 (Singapore: Institute of Southeast Asian Studies).

Slater, J.R. (2000) 'Trends in East Asian Investment in the Transitional Economies'. In P. Artisien-Maximienko (ed.), *Multinationals in Europe* (London: Macmillan).

Thurow, L.C. (1998) 'Asia: The Collapse and the Cure', *New York Review of Books*, 45, 2, 8–11.

Tirado-Angel, I. (1998) 'Overseas Investment and the Development of National Competitiveness in an Envirnonmental Context: Implications for South East Asian Countries'. Unpublished Ph.D. thesis, University of Birmingham.

United Nations, Centre on Transnational Corporations (1992) *World Investment Directory 1992. Volume 1: Asia and the Pacific* (New York: United Nations).

United Nations, Transnational Corporations and Management Division (1992) *World Investment Report 1992*: *Transnational Corporations as Engines of Growth* (New York: United Nations).

United Nations Conference on Trade and Development (1993) *World Investment Report 1993: Transnational Corporations and Integrated International Production* (New York: United Nations).

United Nations Conference on Trade and Development (1994) *World Investment Report 1994*: *Transnational Corporations, Employment and the Workplace*. (New York: United Nations).

United Nations Conference on Trade and Development (1995) *World Investment Report 1995: Transnational Corporations and Competitiveness* (New York: United Nations).

United Nations Conference on Trade and Development (1996a) *World Investment Report 1996: Investment, Trade and International Policy Arrangements* (New York: United Nations).

United Nations Conference on Trade and Development (1996b) *Investing in Asia's Dynamism: European Union Direct Investment in Asia* (New York: United Nations).

United Nations Conference on Trade and Development (1997a) *World Investment Report 1997: Transnational Corporations, Market Structure and Competition Policy* (New York: United Nations).

United Nations Conference on Trade and Development (1997b) *Sharing Asia's Dynamism: Asian Direct Investment in the European Union* (New York and Geneva: United Nations).

United Nations Conference on Trade and Development (1998) *World Investment Report 1998: Trends and Determinants* (New York: United Nations).

Section II
The Asian Financial Crisis

5 Financial Upheaval in the ASEAN Countries: a European Perspective

Françoise Nicolas

INTRODUCTION

After more than a decade of almost uninterrupted strong economic growth, the dynamic ASEAN countries went through a period of heavy financial turbulence from the early summer of 1997. After the European Union (EU) in 1992–93 and Latin America in 1994–95, Southeast Asia was the third region to be hit by a major exchange rate crisis since the beginning of the 1990s. Lessons from the two previous crises may shed light on the mechanisms underlying the most recent developments, even though some features are undoubtedly specific to East Asia thus calling for entirely new analytical approaches. Moreover, because ASEAN countries have been a major engine of growth in the recent past, it appears legitimate to try to assess how the slowdown in economic activity in this region will impact on other regions of the world, in particular on the European Union whose economic recovery in the latter part of the decade was largely export-driven.

The purpose of this chapter is twofold. First, it tries to provide an explanation to the contagion of the financial turbulence throughout Southeast Asia, with particular reference to the European Monetary System (EMS) crisis of 1992–93. Second, the chapter examines the various possible implications of the Asian financial and exchange rate crisis for the performance of individual EU economies as well as for the working of the European Union as an institution. It closes with a conclusion on the scope of future economic cooperation between the European Union and East Asia.

117

THEORETICAL CONSIDERATIONS

One of the leading explanations advanced for almost all exchange rate crises is that the real exchange rate was previously overvalued. As a result, a nominal devaluation occurs to solve the real over-valuation of the currency. This would explain the market speculation against the currencies and the subsequent real devaluation (Godfajn and Valdes, 1996). Traditionally, speculative attacks are thought to occur because governments run macroeconomic policies which appear inconsistent in the longer term with the fixed exchange rate (Flood and Marion, 1996): in other words, attacks are justified because the so-called 'fundamentals' are not right. In his seminal article, Krugman (1979) showed that inconsistencies between domestic economic conditions (fiscal/monetary policy) and an exchange rate commitment will lead to the collapse of a currency peg. The excessive creation of domestic credit due to the monetization of fiscal deficits leads to capital outflows and to a gradual reduction in the government's foreign exchange reserves. Eventually reserves fall to a critical threshold at which a speculative attack is launched, eliminating the authorities' remaining foreign assets. Once reserves are depleted, the exchange rate peg is abandoned and the currency depreciates secularly over time, reflecting the more expansionary stance of policy at home than abroad (Eichengreen *et al.*, 1996). Of course, inconsistencies can be temporarily papered over if the Central Bank has sufficiently large reserves but when these reserves become inadequate, speculators force the issue with a wave of selling (Krugman, 1997). In a nutshell, the so-called 'canonical model' emphasizes the 'natural' and anticipated demise of an inconsistent policy regime. Currency crises are fully predictable in this framework.

The canonical model appeared to fit the experiences of a number of countries in the past: Argentina, Mexico, and Chile in the late 1970s, and France and Italy in the early 1980s. However, the European experience in the early 1990s and the 1994 Mexican peso crisis forced economists to rethink the cause of speculative attacks: many of the European countries, and later Mexico, were running disciplined macroeconomic policies when their currencies were attacked. Sound policy thus cannot apparently guarantee insulation from speculative attack. Yet if inconsistent macroeconomic policies are not the cause, what triggers the attack? A range of so-called 'second generation models' suggests a whole set of possible answers to this question.

Many now believe that a currency crisis can be an unpredictable event, not forced by movements in past or current fundamentals. In contrast to traditional models, 'second generation models' assume that a currency crisis may arise even in the absence of a continuous deterioration in the economic fundamentals, in other words even when macroeconomic policies are consistent with the fixed exchange rate policy. Instead, a spontaneous attack may pull the country off a fixed exchange rate if it brings about a future change in macroeconomic policies. The latter models are based on the possible existence of multiple equilibria (Flood and Marion, 1996), with the switch from one equilibrium to another resulting from changes in expectations. Market participants, while not questioning that current policy is compatible with the indefinite maintenance of the currency peg, anticipate that a successful attack will alter the policy. As a result, it is expected future fundamentals conditional upon an attack taking place, rather than current fundamentals and expected future fundamentals in the absence of an attack, which are incompatible with the peg (Eichengreen *et al.*, 1996). Many of these models thus emphasize the role of self-fulfilling expectations.

In these 'second generation models', the government is thought to choose whether or not to defend a pegged exchange rate by making a trade-off between short-run macroeconomic flexibility and longer-term credibility. The logic of the crisis arises from the fact that defending a parity is more expensive if the market believes that defences will ultimately fail. The important point is that the attack changes the cost–benefit calculation of the government *vis-à-vis* exchange rate policy.[1]

A number of factors other than narrowly defined fundamentals may make the cost of defending the peg too high, thus affecting the choice of policy and the trade-off between credibility and flexibility.[2] While all 'second generation models' assume the exchange rate regime to be vulnerable to shifts in private expectations, they may differ on the role assigned to economic fundamentals. A speculative attack on a currency can develop either as a result of a predicted future deterioration in fundamentals, or purely through self-fulfilling prophecy (Krugman, 1998). The shifts in expectations may have nothing to do with the soundness of domestic economic policy or other market fundamentals but may reflect arbitrary and unpredictable factors. They may also result from any development that can affect the willingness or ability of the government to defend the peg (such as rising unemployment, elections, a fragile financial sector etc.).

Eichengreen *et al.*, (1995) identify a number of mechanisms (not related to economic fundamentals) that may induce a government to abandon the currency peg: situations with economically sound policies which are *politically* unsustainable (the case of France in 1992–93); contagious crises because of competitiveness effects; self-fulfilling speculative attacks (with particular events acting as triggers by coordinating expectations of speculators); and finally strategic explanations. In the context of self-fulfilling attacks, the main challenge is to explain why the speculators all act in the same way and at the same time. But anything may actually set off such self-fulfilling attacks. According to so-called 'sunspot dynamics', any factor becomes relevant if market participants think or believe it is relevant (Krugman, 1997). As a result, focal events may trigger the crisis but herding behaviour may also come into play.

THE EUROPEAN MONETARY SYSTEM CRISIS OF 1992–93

In 1992–93, many European currencies came under attack although the local authorities were running disciplined macroeconomic policies and economic fundamentals were apparently sound. Within the framework of the 'first generation models', there was no obvious reason for France to change its peg to the Deutschemark: the French inflation rate was lower than the German rate, the French current account was in surplus, and the French economy was at the time the only one (apart from Luxembourg) to meet the Maastricht criteria. Yet the desire to fight unemployment appeared inconsistent with the maintenance of the prevailing peg, leading to the expectation that the authorities might choose to shift policy. An application of the 'second generation models' would thus stress the existing French unemployment rate, the evolution of German monetary policy, and the uncertainty over the introduction of European Monetary Union (EMU) as explanations of why agents might have expected a change in French exchange rate policy. The maintenance of a tight monetary policy following the German example was perceived as likely to aggravate unemployment thus pushing up the cost of staying within the EMS, whilst there were rising doubts about the chance of EMU coming into existence as timetabled (after the negative vote in Denmark and the narrow yes vote in France).

The EMS crisis also provided an interesting example of contagion through competitiveness effects.[3] The attack on the Pound Sterling in September 1992 and the subsequent depreciation of the British currency were said to have damaged the international competitiveness of the Republic of Ireland, for which the United Kingdom was the single most important export market, and to have provoked the attack on the punt at the beginning of 1993. By the same token, attacks on Spain in 1992–93 and the depreciation of the peseta were said to have damaged the international competitiveness of Portugal which relied heavily on the Spanish export market, and to have provoked an attack on the escudo despite the virtual absence of imbalances in domestic fundamentals in the country (Eichengreen *et al.*, 1996). The depreciation of the Pound Sterling and the Italian lira may also have affected the French franc. At the very least, these depreciations were perceived to impact upon France's trade and employment situation, thus increasing the pressure on the French Government to abandon its commitment to a close peg to the Deutschemark (Krugman, 1997).

In addition to this competitive devaluation aspect, political commitment to a fixed exchange rate (and the loss of it) may also be subject to herding effects: once Britain and Italy had left the Exchange Rate Mechanism (ERM) in September 1992, it was less costly politically for other countries to devalue or to abandon the peg to the Deutschemark. Because the speculators anticipated this change in the cost of devaluation, they chose to attack these currencies.

THE ASIAN CURRENCY CRISIS OF 1997

While the 1997 financial crisis in Asia was quite different from the EMS crisis in many respects, some commonalities may be highlighted. First, there was the failure of the financial markets to give any weight to the possibility of a crisis until very late. Second, there was the fact that the assumptions of the 'first generation models' were not satisfied. Third, there was the contagion effect. However, the EMS experience sheds light on the process of contagion that engulfed the whole of East Asia in the wake of the collapse of the Thai currency, rather than helping to explain the Thai currency crisis itself.

The Thai crisis certainly does not fit the canonical model: on the eve of the crisis the government was in fiscal balance and was not engaged in irresponsible runaway monetary expansion. Moreover, the savings rate was high and there was no substantial unemployment when the crisis began, despite a marked slowdown in economic growth in 1996. While closer to the pattern described in the 'second generation models', the weaknesses that proved to be lethal in the case of Thailand and which led to speculative attacks were of a fundamentally different nature from those observed in the EMS crisis. In early 1997, the exchange rate commitment was increasingly perceived to be inconsistent with an economic situation characterized by a slowdown in exports, and a loss of competitiveness due to rising wages that were not matched by a rise in productivity. Once the dollar started appreciating *vis-à-vis* the yen and most European currencies (with the exception of the Pound Sterling), the baht peg to the dollar became unsustainable and even more so because of financial excess. Such excess was manifest in particular in the rising share of capital investment flowing not to enhance export promotion in knowledge-intensive or high value-added manufactures and in high technology industries, but in highly speculative and overvalued property ventures financed largely with unhedged short-term borrowing in foreign currency (Sharma, 1998).

Frailties in the banking sector provide a major reason why confidence in the ability of the authorities to maintain the peg was shattered. Financial intermediaries (through the Bangkok International Banking Facilities)[4] were central players in the Thai crisis. They were borrowing short-term money abroad (in dollars) and lending in baht to local investors involved in longer-term projects in particular, but not only, in real estate. As a result, there was a combination of asset-liability and currency mismatches. In addition, the excessive risky lending of these institutions led to an inflation of asset prices and to increases in bad debts. A major feature of the Thai crisis was the fact that a boom-bust cycle in the asset markets preceded the currency crisis,[5] with the downward trend in asset prices starting in late 1995 or early 1996 – see Figure 5.1. At some point the bubble

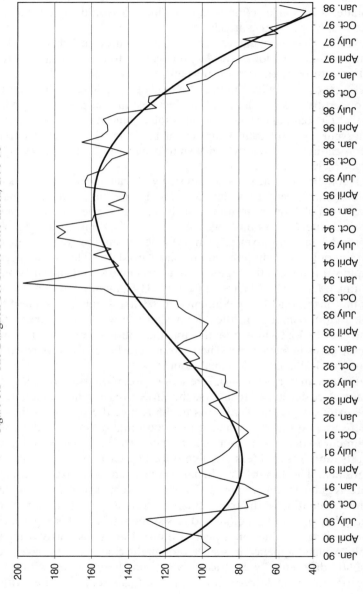

Figure 5.1   The Bangkok Set Share Price Index, 1990–98

Source: Bank of Thailand.

had to burst: real estate prices started to weaken, the value of the collateral of banks plummeted, leading to dwindling confidence and finally speculative attacks.

The foreign exchange reserves started to drop in February 1997, and even more seriously in May in response to repeated speculative attacks. Interest rates were pushed up in response to the attacks, raising the probability that the exchange rate peg would ultimately be dropped because of the resulting increase in non-performing bank loans. The baht appreciated against the dollar after the coordinated intervention by the four Asian central banks in mid-May, but started depreciating in early July and then to collapse once it was allowed to float freely – see Figure 5.2.

If contagion refers to a mechanism by which the currency attack on one country leads to a deterioration in the situation of another country and thus to an attack on its currency, the Asian financial crisis is a perfect example of such a process. By contrast to what was observed in Latin America in 1994–95 when the 'tequila effect' turned out to be both limited and short-lived, the Thai crisis spread rapidly throughout the region at an unexpected speed and with long-lasting effects. The Malaysian Central Bank abandoned the defence of the ringgit on 14 July. Similarly, after first widening the band for the Indonesian rupiah, the Indonesian Central Bank allowed the currency to float a month later, while the South Korean won fell prey to the crisis in November. The currency crash was soon followed by a major GDP contraction in all countries.

Despite major obvious differences between the two episodes, the framework developed to analyse the EMS crisis may help understand the extension of the Thai crisis to the rest of the region. Although there was no formal exchange rate coordination in Asia, the commitment to defend the exchange rate peg was made more and more costly, and less and less credible once one currency, the baht, was devalued. The lack of an official commitment to a fixed exchange rate regime in the region did not help keep speculators at bay. The devaluation of the baht actually served as a focal factor making the other pegs less credible and less sustainable. In this regard, the competitive devaluation aspect of the EMS crisis is partly comparable to what happened in Asia, although the channels may have been slightly different. By contrast to what may be observed within the European Union, intra-regional trade is still extremely limited within

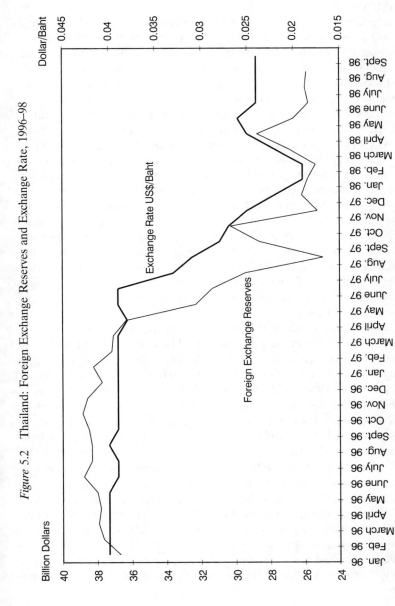

Figure 5.2 Thailand: Foreign Exchange Reserves and Exchange Rate, 1996–98

Source: Bank of Thailand.

ASEAN.[6] ASEAN countries have certainly become steadily more integrated into the wider East Asian economy, but very little of their trade is with other Southeast Asian countries (with the exception of Singapore). Overwhelmingly the trade and investment links are with Japan and the Newly Industrializing Economies (Nicolas, 1997). Thailand's devaluation in July 1997 can be regarded as having had negative repercussions for some of its neighbouring countries, in particular Malaysia and to a lesser extent Indonesia, not so much because of direct trade between the two countries but because their exporters competed in the same third markets (the European Union and Japan). Thailand and Malaysia sell similar products in world export markets. As a result, the devaluation of the baht depressed Malaysian exports and pushed Malaysian authorities to follow suit.

Other channels may further account for the contagion of financial disturbances across countries. Panic and herd behaviour provide another possible explanation. The high-performing Asian economies were in a sense the victims of the 'Asian miracle': belief in such a miracle led to overconfidence, to excessive capital inflows, and to the accumulation of financial imbalances, but once one of the miracle economies was shattered, the overconfidence crumbled and led to excessive outflows.

Finally, the crash of the Thai currency may have also acted as a 'wake-up call',[7] unveiling frailties that had been present all along in neighbouring economies but were systematically ignored by foreign investors. The Thai crisis induced international investors to reassess the creditworthiness of Asian borrowers. While doing so, they found that most of these economies exhibited weaknesses similar to those in Thailand.[8] They had all been affected to varying degrees by an export slowdown in 1996, and they all had current account deficits (though smaller than Thailand's). Moreover, the accumulation of excessive short-term dollar-denominated debt was a common feature of most of the miracle economies. Finally, weaknesses in the financial sector also played a large role in the contagion to the rest of Asia, in particular to Indonesia and Korea.[9] The mechanism underlying the Korean crisis was very similar to the Thai situation, with conventional banks playing the role of the Thai financial intermediaries, and with over-investment in excess capacity and/or in unprofitable manufacturing projects rather than in real estate ventures.

WHAT IS AT STAKE FOR THE EUROPEAN UNION?

The recent financial turbulence in the Southeast Asian countries, and the likely slowdown in their economic activity in the years to come, have raised deep concerns in OECD countries, in particular in the United States and in Japan. In Europe, by contrast, a commonly-held view is that EU economies may be less seriously affected because of their relatively weak economic links with East Asian countries.[10] While there may be some truth in this optimistic point of view, more may yet be at stake for the European Union than is commonly thought, because, among other things, of the implementation of Economic and Monetary Union scheme. Moreover, the Asian crisis may have multiple and to some extent contradictory implications for the European Union.

Trade and Investment Effects

The major channel through which the Asian crisis may be expected to impact on European economies is the trade channel. The possible deterioration of Europe's trade balance with Asia may result both from a rise in Asian exports to the European Union, and from a decline in European exports to Asia. Because European economic recovery has mainly been export-driven so far, European economies appear *a priori* particularly vulnerable to a slowdown in economic activity in Asia. Yet, there seemed to be a consensus that the Asian financial crisis would shave less than half a percentage point off EU's economic growth in 1998.[11] A number of reasons explain why these negative impacts were likely to remain limited, at least in the short run.

Following a sharp currency devaluation, import contraction is usually found to precede export expansion. As a result, a sharp drop in Asian imports from Europe was likely to be observed first. The impact was however expected to remain limited both because the European Union conducts a small share of its trade with Asia[12] – see Table 5.1 – and also because other components of aggregate demand were expected to offset this negative development. The impact of the slowdown should thus vary across sectors and across countries. In this respect, the major trade impact was expected to be for Germany, which was traditionally a large exporter of heavy machinery and capital equipment, the kind of products used in large

Table 5.1　Trade Flows between the Major OECD Countries and Emerging Asia, 1996 (% of exports of exporting country)

Exporting Country/Region	Destination of Exports			
	Five Ailing Economies[1]	China	Other Emerging Asian Economies[2]	Total Emerging Asia
Japan	19.6	5.3	17.5	42.4
Australia and New Zealand[3]	19.9	5.0	12.2	37.1
United States	8.4	1.9	7.9	18.2
European Union[3]	6.5	2.3	6.4	15.2
Canada	1.9	1.1	1.2	4.2

Importing Country/Region	Origin of Imports			
	Five Ailing Economies[1]	China	Other Emerging Asian Economies[2]	Total Emerging Asia
Japan	16.5	11.6	7.1	35.2
Australia and New Zealand[3]	9.2	5.3	8.0	22.5
United States	8.6	6.7	7.6	22.9
European Union[3]	6.5	4.5	6.1	17.1
Canada	2.8	2.1	2.2	7.1

Notes:　(1) The five ailing economies are Korea, Indonesia, Malaysia, the Philippines, and Thailand.
　　　　(2) The other Emerging Asian economies are Hong Kong, Singapore, and Taiwan.
　　　　(3) Excluding intra-regional trade.
Source:　IMF, International Financial Statistics Yearbook; OECD, Direction of Trade Statistics.

government-sponsored development projects that were being cancelled or postponed in the aftermath of the crisis. Yet Italian (and to a lesser extent Spanish) producers also appeared to be severely hit, with a 22 per cent drop in their exports over the period 1996–98, compared to a 10 per cent and a 4 per cent drop for French and German exporters respectively. With regard to sectors, luxury goods were seriously affected by the slowdown in economic activity throughout Asia.

The major fear however was that the EU market would be flooded by cheaply produced goods from Asia, damaging some local industries and leading to lower GDP growth and higher unemployment. Such fears did not materialize, and optimism still prevailed in Europe during the course of 1998. Although devalued currencies should have helped Asian exports regain competitiveness on European markets, this impact remained limited for a number of reasons. Many Asian firms were plagued with liquidity problems in the wake of the financial crisis, and were not thus in a position to take full advantage of the change in their competitive position. The financial intermediation essential for production and external transactions (in particular in the form of export credits) had been seriously weakened in all countries concerned, thus undermining export prospects. In Thailand, in particular, banks faced with liquidity problems were not in a position to extend new credits or to roll over credits to exporting firms. Moreover, the high import content of some of these countries' exports also acted as a brake on export expansion.

Yet there are some indications that Asian exports to Europe were on a rising trend throughout 1998. Italian imports from Asia rose by 30 per cent in the first quarter of 1998, while French and German imports increased by 13 per cent.[13] The European steel industry is a perfect example of a sector that was adversely affected by the Asian crisis. Due to a massive increase in low-priced steel from Asian countries in the wake of the financial crisis, the European Union became a net steel importer for the first time in 1998. In addition to the competitiveness gains induced by the collapse of Asian currencies, the sharp drop in domestic demand for a number of products in Asia pushed down prices in industrial markets and made it difficult for other regions to compete. Such developments will necessarily have far-reaching implications on the labour market in particular, and there is also a real risk that they may rekindle protectionist pressures. As export expansion is usually longer lasting than import contraction in the wake of a currency depreciation, some other impacts of this kind are probably still to come. In the future, however, even if export expansion from ailing Asian countries fully materializes, the impact on the European Union is bound to remain moderate because EU trade with Asia is still quite limited: it represents only 2.3 per cent of the bloc's GDP, while the share of the five most affected Asian countries in EU imports and exports did not exceed 2.5 per cent in 1996 (WTO, 1998).[14]

Although a widening of the EU trade deficit with Asia would be a natural and appropriate adjustment of the economic system to the East Asian developments, there is a risk that mounting protectionist pressures may emerge. The gain in competitiveness by Asian economies (even though it may be short-lived) may affect perceptions and give a new momentum to protectionist pressures in Europe. This is all the more likely if Asian investors also start putting off or giving up investment projects that were planned to come on stream, and create jobs, in Europe. A number of Korean investment projects in Europe have indeed been frozen since the crisis broke out: for example, Daewoo in France, Hyundai in Scotland, and Samsung in England. It should thus be a priority of EU Governments to make sure that they resist the temptation to turn to trade restrictions and choose instead to help the Asian economies through the difficult adjustment period.

While the direct effects on trade and investment flows are likely to remain negligible in the short as well as in the medium term, except maybe in a limited number of narrowly-defined sectors, the indirect effects may be a larger source of concern, although they may be contradictory. These effects may work through various channels. First, asymmetric shocks may make coordination of economic policies as well as the maintenance of economic convergence more difficult, leading to rising difficulties with EMU. Second, and more importantly, the European Union may be negatively affected because of the slowdown in the US rate of growth, triggered by the Asian crisis. Over the past couple of years, world growth has been heavily dependent on the growth of the US economy, making it vulnerable to economic downturns. In this respect, the evolution of the US current account position *vis-à-vis* Asia is particularly worrying. Most of the burden of adjustment resulting from the economic slowdown in Asia has been falling so far on the United States, but the situation may not be sustainable for a protracted period of time, thus leading to spillovers in Europe. If this is the case, because of the rising degree of global interdependence, the European Union will no longer be shielded from the global depression. Third, Japan is also complicating the problem. The bad economic situation in Japan is indeed another matter of concern: the sicker Japan is, the more depreciated the Asian currencies will have to be and the more Asia and Japan will compete in running large trade surpluses against Europe and the United States.

In addition to these possible negative effects which cannot leave EU Member States indifferent to the fate of the five ailing East Asian economies, the crisis may also entail some more positive implications. In some markets, the drop in demand from Asia may exert a positive impact on other regions of the world such as the European Union. Such is the case for the oil market, where prices are being pushed down by sluggish demand, reaching levels unheard of since the pre-first oil shock period. By the same token, the crisis may also provide an opportunity for some countries to get a foothold in the Asian market through the acquisition of majority stakes in Asian firms in distress. Yet, at the time of writing, the expected fire-sale has not really taken place for a number of reasons. On the one hand, there is the reluctance of Asian companies to give away their assets; on the other hand, there is the precarious financial situation of some of the companies that are on sale. Moreover, the bulk of the acquisitions have so far been made either by Asian[15] or by US firms, with European investors clearly lagging behind.

Finance: The Neglected Channel

EU–Asia trade is not the whole story. Another channel through which trouble in Asia might spill over onto European economies is the financial channel. First, because rich countries and developing countries are increasingly linked by financial ties, economic turmoil in Southeast Asia will necessarily impact on the developed world's capital markets. The 'flight to quality' experienced in the wake of the collapse of major East Asian stock markets, carried with it the risk of creating a bubble in the markets of the developed world, and a high degree of volatility was indeed observed on those markets in the course of 1998.

More importantly, casual empirical evidence suggests that European banks are quite heavily involved in the region, to an extent largely inconsistent with the magnitude of their trade involvement. According to the Bank for International Settlements, European banks held some $365 bn in loans outstanding to Asian companies and banks in June 1997, way ahead of their Japanese and US counterparts, with $275 bn and $45 bn respectively.[16] Much of this was in the form of short-term loans. French banks' loans to Asia were equivalent to 2.5 per cent of France's GDP, the ratio was 2.3 per cent for Britain, and 1.8 per cent for Germany, with the highest level reached by Belgium (4 per cent). It should be recognized that the bulk (more than 50 per cent) of these loans were to banks and companies in the relatively solid financial centres of Hong Kong and Singapore. Yet the situation differs across countries, with French banks being relatively more involved in ailing ASEAN economies and South Korea (which accounted for 45 per cent of their total loans to the region) than German and British banks (with 29 per cent and 21 per cent respectively).

Leaving aside the financial centres of Hong Kong and Singapore, the highest internal lending went to South Korea ($103 bn), Thailand ($69 bn), Indonesia ($59 bn), China ($55 bn), Taiwan and Malaysia ($29 bn each) – see Table 5.2. Surprisingly, European banks' exposure to East Asia had clearly increased over the previous two years: their loans accounted for 44 per cent of bank lending to the region by mid-June 1997, compared to 40 per cent only one year earlier. With the exception of the British banks, all European banks heavily increased their exposure to Asian risk in the four years leading up to the crisis.[17] A number of factors may explain this trend. First, there was a perceived need to catch up in a region where European banks had traditionally been less active than their counterparts from the United States and Japan. Second, foreigners' lending behaviour may have been driven by moral hazard considerations because of the tight relationships linking the government, the business, and the banking communities in most East Asian countries. In the belief that the payment of principals and interest on their loans were implicitly guaranteed by the government, foreign banks may not have felt the need to conduct careful analyses of the financial institutions to which they were lending vast amounts of money nor to reduce their exposure even after some of the symptoms of the crisis began to surface

Table 5.2   European Banks' Exposure in Asia, at 30 June 1997

Home Country of Bank	Korea		Thailand		Indonesia		Malaysia		Total	
	$bn	%	$bn	%	$bn	%	$bn	%	$bn	%
Germany	10.3	10.0	7.6	11.0	5.9	10.1	5.8	20.1	29.6	11.4
France	10.3	10.0	4.9	7.1	4.7	8.0	2.9	10.1	22.8	8.8
Great Britain	6.2	6.0	2.8	4.0	4.1	7.0	2.0	6.9	15.1	5.8
Other Europe	9.3	9.0	4.2	6.1	7.6	12.9	2.0	6.9	23.1	8.9
Total Europe	36.2	35.0	19.4	28.0	22.3	38.0	12.7	44.1	90.6	34.8
Japan	23.8	23.0	37.5	54.0	22.9	39.0	10.4	36.1	94.5	36.3
United States	10.3	10.0	4.2	6.1	4.7	8.0	2.3	8.0	21.5	8.3
Other	33.1	32.0	7.6	11.0	8.2	14.0	4.3	14.9	53.3	20.5
Total External Commitments	103.4	100.0	69.4	100.0	58.7	100.0	28.8	100.0	260.3	100.0

Source: Bank for International Settlements.

(Park, 1998). Finally, herding behaviour also played a role, and competitive pressure among foreign investors or fund managers may explain the persistence of lending practices despite rising doubts.[18]

Given this high degree of exposure to Asian risk, European banks will undoubtedly feel the pain. The international credit agency, Standard & Poors, has warned that European banks could face losses of up to $20 bn on their Asian loans. It forecasts that 30 per cent of European bank loans to Thailand, and 50 per cent of their loans to Indonesia, would be non-performing in 1998.[19] As a result, Moody's was also talking of downgrading some European banks (in particular Crédit Lyonnais and Commerzbank) because of their exposure to troubled markets in Asia, while some were put under a credit watch.

It is certainly not easy to get a clear idea of the impact such an exposure will have for European banks in the coming years. In 1997, the impact on profitability remained limited thanks to good growth performances but the situation may easily be reversed because the good performances were based on market activities (as opposed to credit activities), which are by definition highly volatile. French banks are likely to be more vulnerable than others in this respect because of their low margins, in particular compared to their British competitors. While most banks will certainly prove resilient enough to respond to these Asian difficulties and be able to cover increased levels of loan-loss provisioning, the measures they will be forced to take cannot remain completely painless. As a result of their large exposure, all these banks will have to increase their loan provisions in order to protect themselves from potential losses on their Asian operations, thus reducing their earnings and their capacity to lend. This aspect may also take a new dimension, should the crisis extend to other parts of the world. In this respect, the high exposure of German banks in Russia is a major matter of concern. Some other side effects can be expected, such as possible delays in the privatization of State-owned banks such as the French Crédit Lyonnais. Moreover, the resulting increased uncertainty is likely to weigh negatively on the prospects for lending in Europe. What is certain is that European banks are not yet off the hook.

More fundamentally, European banks' high exposure to Asian risks underlines the interdependence between European and Asian economies and the need for a more active role of the new European currency in Asia in the future.

THE ASIAN CRISIS: A MISSED OPPORTUNITY FOR THE EUROPEAN UNION?

Given the high exposure to Asian risk on the part of a number of European financial institutions, it appears all the more surprising that European countries did not play more of a role in the various bailout programmes launched in favour of ailing Asian economies. At the time of the Mexican crisis, it may have been a little more understandable for European countries to let the United States lead the show, but it made far less sense for them to act in the same way in the case of the Asian crisis. Even if the trade impact remains limited and even if European banks are resilient enough to weather the shock, there are good reasons for EU countries to be more active in rescuing ailing Asian economies, be it only to balance the weight of the United States and to help the European Union and the euro gain credibility on the international economic stage. Failure to do so may be a missed opportunity for the European Union. The weak role of the European Union towards developing Asia is also in contradiction with Article 17 of the Maastricht Treaty which sets out the Union's principles for development cooperation, pledging the Community to work for 'the smooth and gradual integration of the developing countries into the world economy' (Bullard *et al.*, 1998: 542). More-over, because the Asian crisis is not restricted to Asia but can be said to be a global crisis (as was exemplified by the extension of financial turbulence to Russia and Latin America in the course of 1998), the European Union cannot afford to be indifferent. This crisis has revealed the inadequacies of markets, governments, and interna-tional institutions in coping with rapid financial liberalization. This provides a good enough reason for the European Union as a bloc to make its voice heard in the debate about how to manage the situation and how to redefine the role of the IMF in a world of global finance.

The involvement of European countries in the rescue packages is however not as limited as may appear at first sight. First because they contribute to the rescue through the IMF, to which they are major contributors: taken together, the EU Member States account for almost 30 per cent of total quotas and thus of total voting rights, compared to 18.25 per cent and 5.67 per cent for the United States and Japan respectively. Moreover, a number of European countries have taken individual initiatives in favour of ailing Asian economies. For instance, during a visit to the region, German Finance Minister Theo Waigel witnessed the signing of an agreement that enabled Germany to provide 150 million marks ($82.4 million) in loan guar-antees to Thai banks and State enterprises to support their purchases of German products or to finance projects involving German firms. Yet the European Union as such did not take any major step in order to support trade flows with Asian countries. As a result, the Euro-pean Union is perceived as paying little attention to the region. This may be a misperception but, in politics, perceptions often matter more than reality, and gestures are important. In mid-February 1998, Thailand and Malaysia urged the European Union to help create a new fund to promote trade and investment in Asia in order to revive Asian economies.[20] The point of this fund was to make liquidity accessible to all countries in the region and to guarantee and encourage investment. The decision made during the Asia–Europe Meeting (ASEM) held in London in early April 1998 to create an ASEM Trust Fund within the World Bank was certainly a step in the right direction, although the amounts involved were restricted to a maximum of 30 million ecu. The purpose of the fund was to finance technical assistance missions to help restructure the financial sector and ease poverty in countries hit by the crisis. It remains to be seen whether this commitment will materialize. It would be a pity, both for Asian countries and the European Union itself, if the Europeans did not live up to their Asian partners' expectations.

The Asian crisis may also be seized by EU Member States as an opportunity to promote the role of the euro as an international reserve currency in Asia and in the world as a whole. Since the rise of the dollar was a major cause of the crisis, there is a strong need for other major currencies to play an anchoring role in the area and, in addition to the yen, there may be a niche for the euro. Moreover, as underlined above, European banks, because of their high exposure to Asian risk, clearly have a vested interest in developing the role of the euro as a trading currency in this part of the world. Beyond the management of the current crisis, enhancing economic cooperation between the two regions and encouraging the European Union to take an active part in helping design the future exchange rate arrangement which is called for in East Asia should be a priority in the coming years. Such a move would certainly be instrumental in making the euro more popular in the region, no doubt a highly desirable achievement for both sides and possibly for the stability of the international monetary system. This is one of the many challenges to be faced by the newly-born euro-zone.

NOTES

1.   Because the government is seen as having the choice to maintain or not maintain the existing peg, the literature on exchange rate escape clauses helps understand in which direction and for what reason the government will move after the attack has taken place.
2.   In the case of Sweden in 1992, the cost of defending the peg became too high, given the existing level of unemployment.
3.   By contrast these competitiveness-related mechanisms played virtually no role in the Mexican crisis.
4.   The BIBF is an offshore centre established in March 1993 with the official aim of turning Thailand into a regional financial centre.
5.   A salient feature of the Asian crisis is the tight link between banking and currency crises. According to Kaminsky and Reinhart (1996), balance of payment crises were unrelated to banking crises during the 1970s when financial markets were highly regulated. Everything changed in the 1980s when banking and balance of payment crises were closely related, with banking crises preceding currency crashes (Dooley, 1997). Krugman (1998) argues further that the Thai crisis does not seem to fit any of

the traditional models of exchange rate crisis, that it was brought about by financial excess and then financial collapse, and concludes that it is a financial crisis rather than an exchange rate crisis.

6. Intra-regional trade accounts for about 20 per cent of total trade; yet because of the specific role played by Singapore, this figure widely overstates the reality.

7. This expression is taken from Goldstein (1998).

8. In the case of Korea, in addition to some similar macroeconomic characteristics, the trigger to the contagion is thought by some to have been the attack on the Hong Kong dollar. As a result of this attack, the South Korean economy suddenly looked vulnerable in the eyes of many investors (Park, 1998).

9. As was the case in Thailand, in both Korea and Indonesia poor prudential supervision and tight relationships between the State, the banks, and the business community led to financial excess based on moral hazard.

10. Such a view is particularly popular in France.

11. All forecasts by major international institutions agree on this point. The European Commission is more optimistic than the OECD, for which the EU growth rate could drop by as much as 0.8 per cent. The IMF calculations also point to a limited negative impact on EU economic growth. The UN Commission for Europe is less optimistic and foresees some difficulties within the European Union because of possible disagreements among the Member States as to the appropriate policies to respond to the crisis.

12. Less than 2.5 per cent of EU merchandize imports came from the five ailing Asian economies in 1996 (WTO). This figure rises to 6.5 per cent if intra-regional trade is excluded.

13. The trade balance between Thailand and France shifted from being in deficit over the previous four years to running a surplus in 1997, as a result of both an expansion of Thai exports to France and a drop in imports from France. This development was felt in the very last part of 1997. The growth of Thai exports to France was a mere 0.6 per cent for the period January–September, and reached 9.1 per cent for the period January–November.

14. Again this share rises to 6.5 per cent if intra-European trade flows are excluded. By contrast, Japan received 16.5 per cent of its imports from these five countries, while the United States buys 8.6 per cent of its merchandise imports from these five countries.

15. Be they local or from neighbouring countries.

16. According to an IMF estimate, by the end of 1996, European banks had lent $318 bn and their Japanese and US counterparts $260 bn and $46 bn respectively in the East Asian countries. The comparison of the figures for December 1996 and June 1997 suggests that European banks increased their involvement throughout the first half of 1997 more resolutely than Japanese and US financial institutions.

17.   French banks increased their loans to South Korean banks by 13 per
      cent in the first half of 1997. Altogether there was a 174 per cent rise in
      the commitment of French banks in the five ailing Asian economies
      over the period 1994–97, a 261 per cent increase in the case of German
      banks, 156 per cent for Dutch banks, and 172 per cent for Belgian
      banks. By contrast, British banks merely increased their exposure by 18
      per cent over the same period (Conseil National du Crédit et du Titre,
      1998).
18.   Ernst-Moritz Lipp, a member of the Board of managing directors of
      Dresdner Bank AG, was quoted as saying 'all banks are under certain
      competitive pressure. If the market is attractive you go with the herd.
      Even if you have doubts you don't stop lending'. *Far Eastern Economic
      Review* (10 February 1998).
19.   'No Pain, no Gain'. *Far Eastern Economic Review* (12 February 1998).
20.   Such a fund would differ from the proposed Asian Monetary Fund,
      which failed to win sufficient support in 1997.

BIBLIOGRAPHY

Agénor, P.R., Bhandari, J.S. and Flood, R.P. (1992) 'Speculative Attacks and
    Models of Balance of Payments Crises', *IMF Staff Papers*, 39, 2.
Bullard, N., Bello, W. and Mallhotra, K. (1998) 'Taming the Tigers: the IMF
    and the Asian Crisis', *Third World Quarterly*, 19, 3, 487–504.
Conseil National du Crédit et du Titre (1998) *Rapport Annuel 1997* (Paris:
    Banque de France).
Dooley, M.P. (1997) 'A Model of Crises in Emerging Markets', NBER
    Working Paper no. 6300.
Edwards, S. (1996) 'The Determinants of the Choice between Fixed and
    Flexible Exchange Rate Regimes', NBER Working Paper no. 5756.
Eichengreen, B., Rose, A.K. and Wyplosz, C. (1995) 'Exchange Market
    Mayhem: the Antecedents and Aftermath of Speculative Attacks', *Eco-
    nomic Policy*, 21, 249–312.
Eichengreen, B., Rose, A.K. and Wyplosz, C. (1996) 'Contagious Currency
    Crises', CEPR Working Paper no. 1453.
Esquivel, G. and Larrain, F. (1998) 'Explaining Currency Crises', mimeo.
Flood, R.P. and Marion, N.P. (1996) 'Speculative Attacks: Fundamentals and
    Self-Fulfilling Prophecies', NBER Working Paper no. 5789.
Goldfajn, I. and Valdes, R. (1996) 'The Aftermath of Appreciations', NBER
    Working Paper no. 5650.
Goldstein, M. (1998) *The Asian Financial Crisis* (Washington DC: Institute
    for International Economics).
Kaminsky, G.L. and Reinhart, C.M. (1996) 'The Twin Crises: the Causes of
    Banking and Balance-of-Payments Problems', International Finance
    Discussion Paper no. 544 (Washington DC: Board of Governors of the
    Federal Reserve System).

Kobayashi, T. (1997) 'The Currency Crisis in Thailand', Fuji Research Paper no. 7 (Fuji Research Institute).

Krugman, P.R. (1979) 'A Model of Balance-of-Payments Crises', *Journal of Money, Credit and Banking*, 11.

Krugman, P.R. (1997) 'Currency Crises'. Paper presented at NBER conference, October.

Krugman, P.R. (1998) 'What Happened to Asia?' Website @web.mit.edu/krugman/www/DISINTER.html

Nicolas, F. (1997) 'Mondialisation et Régionalisation dans les Pays en Développement – les Deux Faces de Janus', *Politique étrangère*, 2, 293–307.

Park, Y.C. (1998) 'The Financial Crisis in Korea: from Miracle to Meltdown', mimeo.

Sachs, J., Tornell, A. and Velasco, A. (1996) 'The Collapse of the Mexican Peso – What Have We Learned?' *Economic Policy*, 22, 13–63.

Sharma, S.D. (1998) 'Asia's Economic Crisis and the IMF', *Survival*, 40, 2, 27–52.

World Trade Organization (1998) website @www.wto.org

6 The Banking Crisis and Competitiveness in the Asian Economies

Diana Hochraich

INTRODUCTION

This chapter will attempt to highlight the underlying mechanisms of the Asian crisis: in addition to the most obvious phenomena, it explores those factors which have long-term effects, such as the impact of capital inflows on the financial and real economies in emerging countries, and the loss of comparative advantage. In particular, it claims that financial instability results from massive capital inflows when these inflows are not accompanied by sound banking system management.

The chapter is structured as follows. In the first section, we present the chain of events leading to the formation of financial bubbles and the subsequent banking crisis. The main flaws of the industrialization model adopted in Asia are presented in the next section, whilst the following section features a brief presentation of the IMF rescue plans and their main long-run consequences. We conclude by describing the changing relations between the Asian developing countries and Europe, the United States, and Japan, and consider the future prospects for Asia.

AN ASSESSMENT OF THE CRISIS

Following deregulation which began in the mid 1980s, the large-scale injection of foreign direct investment (FDI) helped speed up development in many Asian countries. FDI is generally considered to aid progress both by helping to break down financial constraints and by purportedly ushering in technological innovation. These substantial capital injections were accompanied by the rapid growth of credit and of the money supply, resulting in overheating and inflationary pressures which ultimately tended to deteriorate the current account balances of the countries. The injections also have a major impact on stock and property markets: in fact, they undermined the stability of these markets and of the entire financial system.

In the 1990s, many Asian countries were the beneficiaries of foreign capital inflows. As in other regions, official development assistance had dwindled in Asia – see Figure 6.1 – removing what was once a driving-force. Foreign direct investment moved in to take over, followed by portfolio investment whereas commercial credit, which had accounted for significant sums up to the early 1980s, had become practically non-existent.

In the early 1990s, rising demand for credit coming from the overheated economies was met by increased domestic monetary expansion provided by capital inflow. The economies continued to grow too fast, and inflation was rising. Interest rates were raised in order to slow growth and price inflation. However, the interest rates hikes were insufficient to check credit expansion, which was growing more quickly than Gross Domestic Product (GDP). In Thailand and Malaysia, the credit/GDP ratio – see Figure 6.2 – exceeded 100 per cent in 1995 and 1996. In turn, the interest rate differentials – see Figure 6.3 – attracted more foreign capital, most of it volatile (portfolio investments and commercial credit).

In Korea and Thailand, foreign reserves grew substantially, as early as the second half of the 1980s, due to opening up of capital markets. This growth should have led to rising exchange rates. Instead, it was decided that exchange rates would be administratively managed to avoid a drop in the price-competitiveness of Asian exports. In nominal terms, this meant a strong depreciation against the US dollar. In real terms – taking account of inflation – exchange rates seemed to indicate an appreciation of real parities in relation to the US dollar, which in turn steadily depreciated between 1985 and 1995 against other strong currencies, particularly the yen. All in all, trade and current account deficits – see Figure 6.4 – were unavoidable in the face of high investment rates, and more generally, already excessive final demand.

The current account balance was in deficit throughout that period except during the second half of the 1980s. Previously deemed

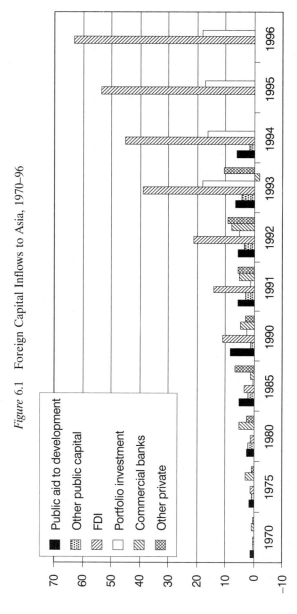

Figure 6.1   Foreign Capital Inflows to Asia, 1970–96

Note: For 1995 and 1996, the only category of capital inflow separately identified is FDI. All other categories are treated as one.

Source: World Bank.

Figure 6.2   The Credit/GDP Ratio for Selected Asian countries, 1985–97

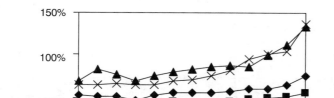

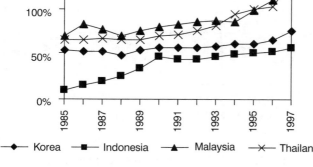

Figure 6.3   Average Money Market Interest Rates in Selected Asian Countries, 1985–96

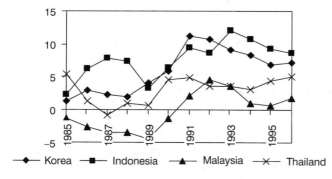

'virtuous' by most economists, resulting from the very substantial investments which were supposed to improve labour productivity in these countries, the assessment of these imbalances has been judged more recently to be disappointing. Despite the high investment rates over the previous decade, industrial competitiveness has not sufficiently improved. On the contrary, the majority of Asian countries face competition from poorer countries in terms of labour costs, while lacking the capacity to sufficiently produce capital-intensive goods. These countries have thus recorded significant slowdowns in

Figure 6.4   The Current Account Balances of Selected Asian Coutries,
1985–96

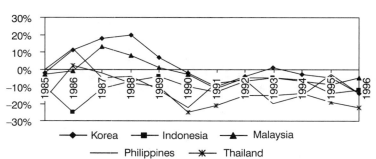

exports. Persistently high deficits have also contributed to loss of confidence in their currencies and to ensuing capital flight.

The loss of external markets, over-investment in unprofitable sectors (property), and domestic economic agents' indebtedness to foreign banks without hedging against exchange rate fluctuations (to avoid excessively high local interest rates), were the ingredients of a crisis which struck nearly all the countries in the region, to varying degrees.

Capital Inflow and Credit Inflation

In Asia's emerging markets, the banks play an essential role as financial intermediaries: 60 per cent of loans in these countries are bank loans. This makes the banking system, which also generally lacks strict prudential oversight, the main target of capital flows' perverse effects. As for the non-banking financial system, the lack of supervision by any monetary authorities leaves it uncontrolled and thus an even greater source of risk.

Increasing foreign reserves created a base for credit extension through larger bank deposits, except when these reserves were used to pay for external debt. In all the region's countries – except Indonesia – increases in the monetary reserves were at least partially obtained through loans from foreign financial institutions (Khan and Reinhardt, 1995). Both these loans and swelling domestic deposits contributed to developing the banking sector.

Two main instruments may be used for sterilization purposes: bank reserve requirements and open market operations. The former pushes up the costs of financial intermediation since interest rates paid on bank reserves are lower than market interest rates. To preserve their profit margins, banks pass this extra cost on to borrowers, who are thus encouraged to seek sources of funding outside the banking system. The latter, the more frequently used indirect sterilization instrument, has a limited impact as open markets are not big enough to absorb all the excess of liquidity in the hands of private agents.

To make the sterilization process more effective, these countries also endeavoured to balance their budgets. Sterilization is not possible so long as existing liquidities can directly or indirectly finance a budget deficit. Most of the countries concerned managed to meet this requirement. But the downside was their governments' inability to finance domestic infrastructure and to provide training for skilled labour so as to improve the technological level of their productive systems.

Thus the expansion of badly allocated bank credit based on capital inflows, at a time when sterilization policies could be circumvented, led to overproduction and excessive current account deficits, making international investors wary. Interest rates hikes could not prevent capital flight: in fact, they only penalized domestic economic agents, who were already faced with liquidity shortages.

Speculation in Real Estate

In almost all the countries, except Korea, the financial crisis was rooted in the real estate sector. In Thailand, the ratio of unoccupied buildings was one of the highest in the region. Lost confidence in the banking sector and currency depreciation compelled foreign banks to stop renewing their revolving credit to Thai banks. Thai short-term debt amounted to $23 billion and non-committed reserves to $7 billion, the equivalent of five to six weeks of imports. In the Philippines, which had a credit to GDP ratio of 60 per cent, the banking system appeared to be more stable than Thailand, where the ratio was over 100 per cent. But bankruptcies of several major firms and over-exposure to real estate had had a weakening effect, even though overcapacity was apparently not as serious as in other countries in the region, limited to 2 per cent of available office floor space. Overall, the situation was somewhat risky: nearly 50 per cent of foreign exchange reserves were made up of portfolio investments and over 40 per cent of securities were held by foreign investors. High interest rates meant to sustain this investment would hamper growth, while devaluation could jeopardize current deficit financing. The Philippines had a large trade deficit equivalent to 8 per cent of GDP, partly offset by remittances from emigrant workers. But as the risk of devaluation increased, these remittances were partly suspended or delayed by emigrant workers wanting to preserve their purchasing power.

In Malaysia, real estate overcapacity was reportedly lower than in Thailand, with vacant office space not exceeding 3 per cent, and construction work having been financed by local funds. Indonesia supplies Japan with raw materials. Despite significant rises in industrial exports, trading in primary commodities allowed it to spread overproduction risks and those of a possible deterioration in terms of trade. Foreign investment is on the decline, and the ecological disaster of 1998 reduced agricultural exports. Political unrest, economic and financial problems all combined to make things worse. Foreign debt, almost as big as Korea's ($135 billion) had been only partially rescheduled. And Indonesian debtors (mostly banks) had failed to service their debts; one member of the government suggested that they wait for 'global renegotiation'. Creditors will most likely have to write off up to 70 per cent of Indonesian debt.

In Korea, overcapacity was specially prevalent in the manufacturing sector where six large conglomerates went bankrupt in the first 10 months of 1997. The large *chaebols* were heavily indebted: their current liabilities represented over 400 per cent of equity. Until recently, these conglomerates could rely on the banks' unconditional assistance to satisfy their cash needs and, though now privatized, Korean banks are still unofficially controlled by government, which continues to encourage credit extension to the *chaebols*. The banking system has become extremely fragile due not only to corporate debt, but also to the same securities being used simultaneously to guarantee several loans, a method also practised by Japanese banks.

Comments

The immediate causes of the crisis were thus: shaky banking systems; overproduction due to excess investment, namely foreign, in almost all the goods produced for export (semiconductors, cars, steel, buildings); real currency appreciation against the yen, resulting in declining competitiveness in the face of Japanese exports; and the speculative bubble in real estate. These factors all contributed to widening current account deficits and growing tensions on foreign exchange markets. Higher interest rates aimed at attracting more capital inflows did not suffice to restore confidence in the currency. This same policy severely hampered the real estate sector, which was already feeling the pinch of shrinking demand. After having lost part of their exchange reserves, the governments failed to avoid currency depreciation.

THE FLAWS IN THE ASIAN MODEL OF INDUSTRIALIZATION

Beyond its cyclical aspects, the crisis has brought to light two major problems: the impossibility of sterilizing foreign capital over the medium-term, and the inadequacy of the technology transfers which the investments were supposed to ensure. These investments have failed to help beneficiaries sufficiently enhance their technological development so as to make up for the losses in cost competitiveness due to wage increases.

The underlying cause of the crisis was thus the industrialization strategy followed by these countries. Its flaws may be summarized by four main characteristics: over-specialization of production, loss of competitiveness, insufficient technology transfer, and an over-reliance on external markets.

Over-Specialization of Production

Industrial production in Asian countries mainly consists of consumer electronics, textiles, and cars. As most production is sold abroad, the same pattern prevails in exports. Value-added in these industrial branches is limited, as most activity is made up of assembly operations. This is very much the case for the electronics industry in Thailand, Malaysia, and China. Here value-added only accounts for 15–17 per cent of total product value, while electronics accounts for more than 30 per cent of exports.

Korea and Taiwan are rather different cases. As products have attained more sophisticated levels, it has become more difficult to upgrade their production methods. The licences they need are no longer public goods, and have become more and more expensive. In spite of the great efforts made to foster national innovation systems, they have yet to achieve technological parity with the industrialized countries. For example, the serious slowdowns recorded in Asian electronics exports in 1996 was due in part to the launching of a new generation of semiconductors by US firms.

Loss of Competitiveness

Contrary to popular belief, wages in many emerging countries have risen much more quickly than productivity. Even allowing for significant errors in measuring productivity levels, most studies arrive at the same qualitative conclusion: low wages combined with still lower productivity result in higher unit labour costs than in the United States. Even when this is not the case, varying productivity levels give rise to substantially diminished differences in unit labour costs. For example, the Korean hourly wage is only 13 per cent of the US wage, but Korean unit labour costs are almost one-half of US costs. Some estimates conclude that, in countries like the Philippines, India, and Malaysia, unit labour costs are even higher than in the United States. If low productivity is well adapted to certain types of production where technological improvement is difficult to introduce (e.g. apparel), it becomes a serious handicap when production switches to more sophisticated processes. High labour costs are inconsistent with low productivity. When labour costs rise, foreign capital flees in search of cheaper labour elsewhere. With their insufficient technology, the former recipient countries have enormous difficulties developing new specializations based on more productive products and processes.

Insufficient Technology Transfer

Technological transfer is often difficult to effect due to the limitations of both the recipient and the technology producing countries. Furthermore, technology transfer is also difficult to measure, as transfer prices mask part of the real value. Economists often assume that FDI may be considered as embodying technology transfer, and thereby to overstate the amounts of those transfers. Many consider only the positive aspects, though some (e.g. Chew and Chew, 1992) show that most research is done in the mother country and very little conveyed to the host country.

For the recipient countries, the difficulties in assimilating technology transfers arise generally from the lack of well-trained labour and from an insufficient R&D infrastructure. Before the current crisis, some social indicators (e.g. rising life expectancy at birth, and decreasing infant mortality everywhere except Indonesia) showed that Asia had improved public health and basic educational standards. Primary school enrolment is substantial, but should be considered in the context of high adult illiteracy particularly in Thailand and Malaysia. Indicators vary considerably from one country to another. High School enrolment is in sharp decline in China, now amounting to merely 50 per cent, down from more than 80 per cent before the reforms. In the other countries, it hardly reaches 50 per cent except in Korea (90 per cent) and Singapore (70 per cent). Other indicators, like female illiteracy which is high everywhere except in Korea, reveal a far from satisfactory situation. As a consequence of the current crisis, all these indicators will register serious setbacks: many Indonesian children have been withdrawn from school, and there are far fewer Korean or Malaysian students in American universities.

Educational standards are generally too low to make best use of technology transfers, and R&D investment efforts are far below the necessary levels to achieve modernization. Even in countries like Korea, where investment rates approach those (around 2 per cent of GDP) in industrialized countries, this effort has been undertaken only over the past few years, whilst building a national innovation system takes at least half a century! Besides, the bill must be paid with real money. Certainly Korean firms have earned their technology credentials, holding about 8.8 per cent of US patents. But half of these patents are adaptations or modifications of existing production processes, whilst only 10 per cent involve key components. The Korean R&D capital stock amounts to around 1.5 per cent that of the United States and 6 per cent that of Japan. This capital is concentrated in very few industries, and in very few firms. Almost 30 per cent of the R&D capital stock is owned by the five top firms, compared with only 17 per cent in the United States and 18 per cent in Japan.

In the industrialized countries, the companies that possess technology are most often oligopolies, and behave as such. They introduce entry barriers, and transfer knowledge only when it becomes a public good. Scientific and technological knowledge are in most cases strictly kept under the control of multinational corporations. As parent establishments impose final prices on subcontracting companies and subsidiaries, and the biggest share of the profits is obtained from distribution, this leaves little ground for R&D investment in countries where most firms are linked to multinationals as subcontractors.

THE IMF RESCUE PACKAGES

Between August and December 1997, three countries – Thailand, Indonesia, and Korea – were on the brink of payment default and turned to the IMF for bailouts. Rescue packages amounted to unprecedented sums: $17 bn for Thailand, $43 bn for Indonesia, and $57 bn for Korea. Not all the money came out of IMF's pockets, of course. Other international financial institutions – the World Bank, the Asian Development Bank, and the Bank of International Settlements – also participated. Individual countries, such as Japan, the United States, Singapore, and even Australia, contributed as well.

The IMF aid was not sufficient to head off a serious credit crunch, as the funds were meant to rebuild monetary reserves rather than to reimburse foreign creditors. As a consequence, Korea found itself once again on the brink of payment default. Korea was eventually rescued by its creditors' banks on the advice of the US Government, which feared the social and political consequences of Korea's bankruptcy. Just $25 bn of the total Korean short-term debt, amounting to $92 bn, was rescheduled. This debt was to be reimbursed in one to three years, guaranteed by the Korean Government.

The draconian conditions which accompanied the bailouts contributed to a worsening of the crisis in the first months of implementation of IMF plans. First, the IMF demanded that failed banks and financial institutions be left to go bankrupt in the three countries. In Thailand, for instance, the IMF demanded the closure of most of the 58 failed financial companies into which the Thai Government had poured approximately US$19 bn since the beginning of the crisis, slightly over 10 per cent of GDP. In Indonesia, 16 banks were closed for the same reasons, while Korea intended to shut down many of its failed merchant banks (i.e. those which finance small and medium enterprises). As many economists have pointed out, this measure only aggravated the banking crisis.

Second, the IMF 'advised' that bank capital itself must be more open to foreign investors. Until then, foreign capital ownership of Thai banks has been limited to 25 per cent. This was also true for productive capital. In Korea, the IMF required that the *chaebols* allow up to 50 per cent foreign investment in their capital – against 26 per cent previously – and this ratio will be 55 per cent in 1999. Bank mergers were also encouraged.

Other measures were designed to increase government revenue to ensure a balanced budget, which would not be reached for the first time in ten years. In Thailand, for example, the IMF demanded increased value-added taxes (VAT) and public spending cutbacks. In Indonesia, it called for reductions in fuel and staple goods subsidies, in spite of sky-rocketing prices and hunger riots. These measures resulted in popular rioting, and the ousting of the former President (Indonesia), strikes (Korea), and social unrest (Malaysia). While alone in not seeking an IMF bailout, Malaysia unilaterally introduced similar austerity measures.

These measures were accompanied by steep interest rate hikes aimed at propping up currencies, but resulting as well in a tight credit crunch. Such measures thus helped to kill off companies which otherwise might have survived. The IMF recipes were considered appropriate for the Latin American debt crisis in the 1980s when huge fiscal deficits were the rule, but not to Asian countries under the conditions in the late 1990s where deficits were not the case.

As a consequence of this policy mix, all the concerned countries entered deep recessions in early 1998 which contributed to switching from current account deficits to surpluses. Meanwhile, their international reserves grew bigger and their exchange rates stabilized, but only after the second quarter of the same year. Taking account of these results, the IMF softened its recommendations. Interest rates were lowered, fiscal deficits allowed, and Indonesian debt was partially rescheduled. But the damage done by the former policy mix was deep: the recession had provoked still higher rates of bad debts, representing 35 per cent of total credit in Korea and 60 per cent in Indonesia. Easing interest rates was useless in this case: due to weaknesses in their balance sheet structure, the banks were unable to extend more credit. Furthermore, private demand had collapsed and public demand was not strong enough to pull these countries out of recession. There was therefore no more incentive to request new credit lines. For instance, while Malaysia's credit growth target was 8 per cent, there was not enough demand to meet it. In short, whilst the IMF plans were designed to restructure the banking systems, the measures they introduced further aggravated the banking crisis.

CRITICISMS OF THE CONDITIONS ATTACHED TO THE IMF RESCUE PACKAGES

The first objection came from the Asian countries themselves, as IMF conditionality was considered to be a challenge to national sovereignty and generally implied a sharp fall in the living standards of the working classes. Calls to close failed financial institutions and industrial companies will harm the vested interests of leading classes and, for this reason, it is likely that such measures will be very incompletely applied, at best. The only possibility which might bring about the comprehensive restructuring in these countries would be to open up their capital to foreign investors, who have the means to effect the restructuring. Local criticism also targeted the fickleness of creditors: the acknowledgement that current difficulties were not only due to imprudent management but also to a very brisk change in market expectations. When the Asian countries 'all of a sudden' became unreliable, foreign banks which had been lending for years refused to renew their rolling credits. The credit crunch followed and triggered off the payments shortfall.

The second major objection came from US Senators, who questioned the legitimacy of the bailouts. The Senators charged that bailouts don't protect the concerned countries, but rather their Western lenders. Yet, according to liberal principles, Western banks that lent a huge amount of money without securing payment guarantees must go bankrupt too. These lenders relied on the 'moral hazard' assumption that, if the countries didn't want to pay back their loans, international rescue plans would force their hand to do so! But more and more Americans feel that they should not have to pay for imprudent business practices. This explains why the Administration encountered stiff opposition when soliciting Senate approval for an additional $18 bn in US contributions to enlarge IMF capital.

For IMF officials the only way to avoid the undesired effects of speculation was to prevent speculators' negative expectations. Accordingly, high interest rates were supposed to maintain foreign capital, whilst deep restructuring in banking and the productive sectors should restore confidence in the economic fundamentals and spark new capital influx. This policy was carried out at great cost and with questionable results. Currencies were stabilized and foreign reserves grew, but any new accident anywhere – in China, Hong Kong, Brazil or, as was the case with the Russian markets in Summer 1998 – could set off a new wave of capital flight. The Russian debacle is what destabilized the Hong Kong market in August 1998.

High interest rates have proven incapable of preventing foreign capital flight, as the Brazilian case proves again. Since the beginning of the Asian crisis, Brazil's short-term interest rates fluctuated between 20 and 40 per cent. But in August, following the Russian crisis, Brazil lost $35 bn, as much as it gained from privatizing its mining companies and public services. In the meantime, as interest rates rose, the already huge Brazilian public debt, accumulated from annual fiscal deficits, sky-rocketed. To restore market confidence in the Brazilian economy would have meant implementing even more severe austerity policies, cutting back fiscal expenditures, and raising taxes. This would have resulted in one of the worst recessions that Brazil had known. The wiser alternative for Brazil, as for all other countries, would be to introduce exchange controls as soon as speculative attacks begin. Malaysia did, but too late and, as the only country having done so, it exposed itself to tough retaliation measures.

The short-run restructuring measures have, so far, been rather marginal and ineffective. In Korea, the closing of five banks only affected 5 per cent of deposits, while the five biggest *chaebols*, which monopolize most of the outstanding credit, refused to restructure their companies. In Indonesia and Thailand, the Special Commissions appointed to clean up the banking system were thwarted by the Governments' protection of the former system. In the long-run, public expenditure restrictions would result in lower educational standards, less R&D, and poorer infrastructure, making it even more difficult to overcome these countries' shortfalls in national competitiveness. The IMF continues to advocate deregulated financial markets even in the face of recent events. Yet the crisis seems to show that Asian countries are not ready to fully open up.

THE FUTURE PROSPECTS FOR ASIA

The severe recession which the Asian countries have experienced since 1998, together with the urge to solve the crisis, could instigate durable changes in these countries' economic structures. The banking systems need to be reformed and firms restructured, including sharp reductions (nearly 60 per cent in most cases) in production capacity. This would necessarily involve the growing participation of foreign firms. Local businessmen will not be eager to participate in such radical changes which would undermine their economic and political command. On the other hand, foreign businessmen don't seem that eager to take over these countries' assets, either in production companies or in banks. To date, very few banks and factories have been sold to foreign firms. And whatever the progress made in introducing reforms, growth will resume at a much slower pace for two reasons. First, restructuring will take a long time, as the Japanese example shows. Second, the days of easy money flowing into Asia are over.

The prospects are somewhat different in Korea, Taiwan, and Singapore, than in the ASEAN countries. In the former, transnational corporations (mostly American) may associate with local industrial companies to impose their rule. In the latter where technology is too backward, penetration will most likely take place through banking systems. US banks will try to update management and accounting systems. The penetration process was already underway at the end of 1998 as the Thai Government launched an auction to sell off $24 bn in loans, belonging to the 56 failed finance companies closed in 1997. These companies' assets were mainly offered to foreign banks at a discount of about 60 per cent, but the auction results were disappointing and very few assets were bought. Thailand is the first country to attempt such a measure, and it ran into strong opposition from nationalists, even within the Government.

Many other consequences will follow. Japan has been hard hit by the crisis, especially through its banking system which was highly exposed to Asian risk and stricken by a domestic credit crisis. Many manufacturing companies are quitting their operations in Asian countries, or scaling down their operations after enduring huge losses. One upshot of this crisis could be a declining Japanese presence in the region.

China and Hong Kong have been so far spared by the depreciation wave, but have met many difficulties rooted in the crisis nonetheless. The Hong Kong dollar was under serious attack: the banking system in the Special Administrative Region has been exposed to other Asian countries' risk, mainly China, Indonesia, and Thailand. To defend the Hong Kong dollar peg to the US dollar, interest rates were raised, thus weakening the real estate and banking industries and driving the stock market to historical lows. Sustaining the peg will mean re-adjusting the real economy through falls in wages and employment rates, thus inducing sharp drops in both the growth rate and prices.

China might appear to be protected against capital flight as its currency is not convertible. But two arguments tend to suggest the contrary. First, residents may – under certain conditions – hold bank accounts in hard currency. So they could withdraw deposits from public banks, exchange yuan for dollars on the black market, and put them into foreign banks that have Mainland or Hong Kong affiliates. Second, China has been experiencing huge capital outflows through Hong Kong, through both formal and informal channels, for 20 years. Indeed this capital has contributed largely to the rise in the Hong Kong stock and residential markets. In 1998, capital flight was evaluated at more than $20 bn. So far, there has been no need for depreciation of the yuan to ensure price competitiveness as real wages are still lower in China than in neighbouring countries, and prices are falling. But China inevitably faces cut-throat competition from its Asian partners. The Asian crisis is making domestic problems harder to solve, and China is hard pressed to restructure its public sector. This will become an increasingly difficult task as the Asian crisis results in reduced FDI and capital inflows, which would help to smooth the social and economic consequences of restructuring.

For the rest of the world, the crisis has created contradictory effects, both opportunities and threats. For the United States, recent events have reasserted its political and financial supremacy, confirming its role as a 'lender of last resort' in cases of financial crisis, even if direct intervention follows the IMF's lead. Heavily indebted in dollars, the Asian countries' demand strengthens the dollar's place as a reserve currency. For American firms, the crisis presents a new opportunity to return to Asia after 10 years of unsuccessful attempts. With the lifting of restrictions on domestic company capital, both in banking and in manufacturing, US multi-nationals will challenge or even replace Japanese interests. And yet these business opportunities may only partially outweigh (or unequally reflect) the trade imbalances, fluctuating financial markets, and downward pressures on the US economy resulting from the Asian readjustment.

The European Union has done little to implement crisis management policies. It has made few, if any, significant proposals and their financial support ($5 bn, on top of their participation in IMF packages) was minor. As noted earlier in this volume, the European Union and Asian countries have been been trying to set up a framework of mutual relations, whose most obvious manifestation has been the two Asia–Europe Meetings (ASEM). The first in 1996 witnessed a substantial consensus around common targets regarding such issues as investment regimes, tariff and trade barriers, and reinforcement of multilateral agreements. In the second meeting in 1998, expectations were quite different on both sides. Already stricken by crisis, the Asian countries expected some substantial financial aid from their European counterparts, while the latter were only willing to offer some technical advice (mainly on banking management and supervision) and small amounts of money (£5 m from the UK Government, 50 m francs from the French Government, and 15 m ecu from the European Union). The meeting proved to be rather disappointing for the Asian countries.

As the crisis continues, economic relations between Asian and European countries may distend. Indeed, the huge contraction of Asian markets (almost 20 per cent) makes it is less profitable for European countries to expand their businesses. European banks have already undergone massive losses due to heavy exposure to Asian risk. Indeed, European banks have proved to be much more exposed than their American or Japanese colleagues – see the discussion in Chapter 5.

The prospects of the euro as a reserve currency are also at stake. As Asian countries are heavily indebted in dollars, they may prefer to continue stocking most of their reserves in dollars. A few weeks before the euro was launched, European bankers and officials were trying to convince Asian governments to convert part of their reserves into euros. A strong euro, they contended, would open new opportunities for Asian trade and investment with Europe. But before taking any decision, Asian officials wanted to judge the euro's stability as a reserve currency. For them, this is more important than any savings made by using euros for trade and investment transactions in Europe. This is not exactly the view of the Asian Development Bank, which stated in a recent report that the region's central banks would have to switch some of their reserves away from dollars, if only to reflect trade patterns more accurately. Asian trade with Europe amounts to around 15 per cent of its total trade, whilst approximately 12 per cent of its reserves are currently held in European currencies.

WHAT FUTURE MODEL FOR DEVELOPING COUNTRIES?

In the 1980s, when the debt crisis broke out in Latin America, the IMF declared the failure of the import-substitution model of economic development. The export-led development model was introduced. During the ensuing 10 years, the International Monetary authorities have been presenting the Asian countries which have applied this model as examples for all developing countries, despite their lack of democracy and transparency. Now that the underpinnings of this model have come to light, the IMF criticizes it, arguing that its implementation was 'skewed'. But the current crisis is the natural outcome of the export-led growth model (Hochraich, 1998). The International Monetary authorities are now calling for more openness, more transparency, and more implementation of Western methods and Western rules. As in Latin America after their debt crisis, tentative restructuring has resulted in a protracted recession, together with a steep fall in the investment ratio. Under these conditions, there is no reason to suppose that Asian countries will continue to catch up with the industrialized countries. Latin America then endured what has been referred to as a 'lost decade'. Why should Asia be different?

BIBLIOGRAPHY

Amsden, A. (1989) *Asia's Next Giant* (New York: Oxford University Press).

Chew, S. and Chew, R. (1992) 'Technology Transfer from Japan to ASEAN: Trends and Prospects'. In S. Tokunaga (ed.), *Japan's Foreign Investment and Asian Economic Interdependence*, pp. 11–134 (Tokyo: University of Tokyo Press).

Collins, S. and Bosworth, B. (1996) 'Economic Growth in East Asia: Accumulation versus Assimilation', *Brookings Papers on Economic Activity*, 2.

Goldstein, M. and Turner, P. (1996) 'Banking Crises in Emerging Economies: Origins and Policy Options', BIS Economic Papers no. 46 (Basle: Bank for International Settlements).

Hochraich, D. (1998) 'Globalisation de la Production et Industrialisation par les Exportations: une Sortie du Sous-Développement?' *Mondialisation et Régionalisation des économies*, no. 27.

Khan, M. and Reinhardt, C. (1995) 'Capital Flows in the APEC Region', IMF Occasional Paper no. 122 (Washington DC: IMF).

Lee, Jaimin (1995) 'Comparative Advantage in Manufacturing as a Determinant of Industrialization: the Korean Case', *World Development*, 23, 7, 1195–214.

Turner, P. and Van't dack, J. (1993) 'Measuring International Price and Cost Competitiveness', BIS Economic Papers no. 39 (Basle: Bank for International Settlements).

Young, A. (1994) 'Lessons from the East Asian NECS: A Contratian View', *European Economic Review*, 38, 963–73.

Section III
Microeconomic Perspectives

7 Floating Market: Currency Crises and ASEAN Tourism Exports

Jim Newton

INTRODUCTION

The objective of this chapter is to explore the relationship between the Asian Financial Crisis of 1997 and ASEAN tourism exports. The currency flotations that began in 1997 led to falling export prices for both goods and services, especially in Malaysia, Indonesia, and Thailand. However increases in merchandise exports, in both volume and value terms, were constrained by financing problems, including not only liquidity difficulties but also foreign exchange shortages that restricted the import of the intermediate goods needed for the production of manufactures. The tourism industries of the affected countries did not face these constraints. On the contrary, the lower exchange rates should give rise to lower prices for hotels, food, entertainment, and shopping, and thus provide an effective stimulus to growth. Consequently a deliberate policy of expanding tourism exports would appear to offer a feasible solution to the economic problems that are the legacy of the crisis.

This policy choice has been advanced most strongly by tourism industry groups and has found at least rhetorical support amongst Asian governments. The anticipated benefits include contributions to the current account, to the foreign exchange reserves, to domestic output, and to employment. In short, tourism was seen as a path to an export-led recovery for the crisis-hit ASEAN economies. In order to review this prescription, this chapter begins with a review of the global tourism industry, before considering the dimensions of the industry in the ASEAN countries. The example of Thailand is then considered in greater detail.

TOURISM: A GLOBAL OVERVIEW

The propensity of much tourism analysis to focus on developing countries tends to obscure the fact that tourism is a much larger activity in the industrial countries than in the periphery or semi-periphery. The generally accepted, although somewhat crude, measure of tourism activity is the number of tourist arrivals. Global arrivals amounted to 592 million in 1996, an increase of 4.5 per cent over 1995. Seventy per cent of the total international tourism business was accounted for by visitors to the OECD countries.[1] The world's leading destination overall was Europe, with almost 60 per cent of total arrivals. The European Union received 239 million tourists, or 40 per cent of the world total. The world's most visited country was France, with over 61 million arrivals in 1996, followed by the United States, Spain, and Italy.

The economic impact of tourism is, of course, better judged by financial measures.[2] In 1996, the total expenditure of international tourists was estimated to be US$423 billion, 7.6 per cent higher than a year before. In the ten years from 1986, this measure had trebled. The industrialized countries captured the greatest share of the revenues. The OECD countries were the major recipients, with over 70 per cent of revenues. The European Union's share was 39 per cent or US$165 billion. Whilst the United States was not the most visited country, with less than 8 per cent of total arrivals, it nevertheless received the highest share of tourist expenditures. US receipts from tourism amounted to US$64 billion in 1996, over 15 per cent of the global total and more than double the amount received by Spain, the next highest earner.

The countries of East Asia and the Pacific (EAP) are the next most significant regional destinations for tourists after Europe and North America. In 1996 the EAP countries received 90 million tourists and US$82 billion in revenues. Whilst considerably less than Europe at present, growth has exceeded any other region for nearly two decades, with the share of both arrivals and receipts roughly quadrupling since 1975.

ASEAN countries received roughly one-third of the total Asian arrivals and, with US$31 billion in receipts, almost 40 per cent of total Asian revenues. Compared to both world and EU averages, growth in ASEAN has been rapid during the 1990s. From 1990 to 1996, arrivals rose on average by over 5 per cent per year, double that in the European Union. Receipts rose even more significantly, by 14 per cent per year on average, nearly three times the EU rate and the highest of any regional grouping measured by the World Tourism Organization.[3]

Thus tourism in ASEAN is a significant and rapidly growing export business. As Hitchcock *et al.* (1993: 1) comment, 'tourism has become one of South-East Asia's foremost industries . . . the members of the Association of South-East Asian Nations (ASEAN) are experiencing a boom in both foreign and domestic tourism. The number of foreign visitors has doubled, receipts from tourism have tripled during the last decade, and tourism has become the leading source of foreign exchange in countries like Thailand. Tourism is the second largest industry in the Philippines and the third largest earner of foreign currency in Singapore. In Indonesia tourism has moved into fourth place, outstripping rubber and coffee as an earner of foreign exchange in 1990'. Table 7.1 shows a recent profile of tourism arrivals to ASEAN.[4] Vietnam has shown the most rapid growth over the five years to 1996, although from a low base, and still lags well behind the other members. The current leaders in the arrivals league, by a wide margin, are Singapore and Thailand, which suggests that the industry plays a more significant part overall there than elsewhere in ASEAN.

Table 7.2 shows receipts from tourist expenditures for the main ASEAN tourist destinations. Thailand is the principal recipient with an estimated US$7.4 billion in 1995, whilst Indonesia claims the second highest inflows.

Table 7.1   Visitor Arrivals to ASEAN Countries, 1992–96

	1992	*1993*	*1994*	*1995*	*1996*
Indonesia[1]	3,064,161	3,403,138	4,006,312	4,324,229	n.a.
Malaysia	2,345,823	2,499,274	2,753,430	2,991,210	2,952,856
Philippines	1,157,241	1,372,097	1,573,821	1,760,163	2,050,117
Singapore	5,989,940	6,425,778	6,898,951	7,137,255	7,298,592
Thailand	5,136,443	5,760,533	6,166,496	6,923,384	7,603,702
Vietnam	440,000	601,527	940,707	1,358,182	1,410,248
Total	**18,133,608**	**20,062,347**	**22,339,717**	**24,494,423**	**21,315,515**

Note: (1) No data reported for Indonesia for 1996.
Source: Pacific Asia Travel Association (PATA), *1996 Annual Statistical Report.*

Table 7.2   ASEAN Tourism Receipts, 1993–95

	1993	1994	1995
Indonesia	3,987,559	4,785,259	5,228,340
Malaysia	1,899,700	3,194,500	3,626,450
Philippines	2,122,300	2,479,560	2,453,960
Singapore	3,412,079	3,342,500	3,860,100
Thailand	5,013,800	5,762,340	7,400,000

Source: Pacific Asia Travel Association (PATA), *1996 Annual Statistical Report.*

THAILAND: A CASE STUDY OF TOURISM IN ASEAN

These broad figures suggest, therefore, that tourism plays a more significant role in Thailand's economy than in the other members of ASEAN, although it is also of some considerable importance in other member countries. In common with the other ASEAN members, Thailand receives the greatest share of its arrivals from East Asia and from other ASEAN countries. This is only to be expected, given the physical proximity of these short-haul markets. To a large extent, the exceptional growth rates recorded in Asian destinations were due to rising intra-Asian tourism as these economies expanded in the 1970s and 1980s. Europeans, however, play a greater role in the composition of Thailand's tourist arrivals than elsewhere, as Table 7.3 shows.

Table 7.3   European Visitor Arrivals to ASEAN Countries, 1992–96

	1992	1993	1994	1995	1996
Indonesia[1]	557,950	634,196	791,657	786,610	n.a.
Malaysia	309,010	336,348	367,328	379,413	386,582
Philippines	148,968	177,181	202,545	225,602	266,869
Singapore	939,143	995,994	985,698	940,414	976,761
Thailand	1,306,027	1,411,057	1,509,483	1,509,198	1,694,600
Vietnam	25,866	64,959	133,560	171,019	114,291
Total	**3,286,964**	**3,619,735**	**3,990,271**	**4,012,256**	**3,439,103**

Note: (1) No data reported for Indonesia for 1996.
Source: Pacific Asia Travel Association (PATA), *1996 Annual Statistical Report.*

Table 7.4 Tourist Nights of International Tourist
Arrivals to Thailand, 1995–96

	1995	1996
Total	52,375,093	59,212,007
From East Asia	21,342,854	24,605,937
From Europe	20,784,375	22,016,500
From the Americas	3,676,260	4,397,716

Source: Tourism Authority of Thailand, *Statistical
Report 1996.*

In 1996, over 70 per cent more Europeans visited Thailand than Singapore, the next most popular country.

The European market is significant for Thailand in another way also. Europeans stay longer than visitors from the short-haul Asian markets and also than those from the other major long-haul market, North America. It is likely that a significant proportion of the European visitors is made up of budget travellers, or 'backpackers', who stay longer than the average, hence increasing the total count. However, official data do not provide this breakdown and it is not possible to estimate the influence of this group. Table 7.4 shows the total number of nights spent in Thailand by visitors from each of the main source regions. By this measure, the European market becomes almost as significant as the Asian. This observation has important implications, given the concern with tourism's role as a potential locomotive to help haul the Thai economy out of the current crisis.

Thailand emerges from this brief review as a logical subject for more detailed analysis. Its tourism industry is the largest in ASEAN both in terms of visitor arrivals and in terms of tourism revenues, and it has the largest share of the European market, a share that is magnified greatly if total stay is considered rather than tourist arrivals alone. Furthermore, there is an additional important reason for studying the Thai case in greater depth, which is based on events that occurred during the late 1980s. Tourism in Thailand rose dramatically in 1986, 1987, and 1988. This increase has been attributed, at least in part, to a marketing campaign known as 'Visit Thailand Year 1987'. If Thailand is to expand tourism to help overcome the economic problems of the late 1990s, policy and strategy can perhaps be based on the experience of the past. In other words, Thailand did it before so can it be done again?

Addressing this question requires a more detailed examination of the potential economic effects of tourism. In the short term, turning to tourism as an easy remedy for economic ills can appear an attractive option. In the immediate aftermath of Thailand's devaluation, the government's response was to latch onto tourism as the natural engine of recovery.[5] The reasons for this continued. There was an on-going liquidity crisis in Thailand[6] that, even some ten months after the initial devaluation, constrained both the export of goods in the short-term and any further expansion of manufacturing plant to take longer-term advantage of the devalued baht. As tourism does not face these same constraints, it was not surprising that it was seen as a ready solution – a quick fix, perhaps, for an economy in trouble.

Reinforcing such official views have been the pronouncements of various international tourism lobby groups. One such group is the World Travel and Tourism Council (WTTC), whose members include the senior executives of the world's major airlines, hotel groups, and tour operators. The WTTC has promulgated a 'Millennium Vision' for tourism, which urges Asian governments to seize the opportunities presented by the new currency parities. The Council claims that the 'current economic problems... offer a real opportunity to capitalise on the comparative advantages of tourism as the area's fastest growing foreign exchange earner' and vows that it 'stands ready to help this endeavour in any and every way possible'.[7] More recently, the WTTC announced[8] that it would conduct an intensive study of the Thai travel and tourism industry, to determine the full impact of the industry on the economy and employment.[9]

THE ECONOMIC IMPACT OF TOURISM

Sinclair and Vokes (1993) provide a useful sketch of the main macroeconomic categories in which tourism may have an impact. They identify three possible contributions: to the balance of payments, to income generation, and to employment. One of the foremost justifications for promoting tourism development, especially in lower income countries, has been the potential for increasing receipts of foreign currency. However, it is recognized, though less widely discussed, that currency outflows also occur, particularly to pay for tourism-related imports of both goods and services. A further issue is the possible instability of tourism earnings, given the sensitivity of tourism to external shocks and to security-related issues.

Much of the work on the impact of tourism on national income has focused on the multiplier effect, using either Keynesian models or input–output tables. Both methodologies attempt to take account of leakages from the circular flow, especially import leakages, and both recognize that the total impact on GDP is much greater than the nominal value of tourism receipts. Empirical work is, however, rather limited and has concentrated on countries that lend themselves to relatively easy analysis. The studies cited by Sinclair and Vokes (1993) are principally of small island economies, particularly in the Pacific, together with studies of the city-states of Hong Kong and Singapore, and only one of a geographically larger nation, the Philippines.

The same review catalogues the limited amount of serious academic work on the issue of employment creation. Again there is the notion of both indirect as well as direct effects, but with little in the way of rigorous, or even non-rigorous, methodologies. The authors state (1993: 206), perhaps tellingly, that tourism 'is often assumed to be advantageous because of its employment-creation potential'. Tourism is taken to be a labour-intensive industry, and hence this should validate the assumption. However, the authors also point out that certain aspects of the industry are highly capital-intensive, especially when the provision of airport and road infrastructure is considered.

Thailand was facing problems in each of these three areas in the late 1990s. GDP was forecast to grow at a much lower rate than before the devaluation. For many reasons, not the least of which were the liquidity crunch and the associated high lending rates, bankruptcies were increasing and existing businesses were laying off staff, thus increasing unemployment rates. Many of the newly unemployed were returning to their rural roots, to which they had previously remitted part of their earnings to support their families and in which there was no prospect for wage earning.[10] The social consequences of this could be profound. The external account on trade was returning to balance, but this was almost wholly as a consequence of falling imports and not because of renewed export competitiveness. The seriousness of the balance of payments situation was underlined by the Thai Prime Minister at the ASEM meeting in April 1998 when he stated that exports in January 1998 were 8 per cent down on the same month in the previous year.[11] One of the tasks facing the Thai Government was the rebuilding of the reserves and, in this, tourism's putative ability to generate foreign exchange could therefore be of value.

However, the depreciation of the baht pulled simultaneously in two opposite directions. The exchange rate trajectory from early July 1997 revealed a steady decline from roughly 25 baht to the US dollar, to a low of almost 57 baht in mid-January 1998, followed by a recovery to of around 40 baht by the end of 1998. This amounted to an effective discount of 60 per cent on all purchases denominated in baht, which would last until domestic inflation or attempts by local businesses to seize the windfall profits eliminated the discount. On the one hand, this might stimulate an increase in the numbers of arrivals and possibly also the length of stay. On the other hand, Thailand will have to boost earnings in baht by a corresponding 60 per cent to earn the same amount of foreign exchange in 1998 compared to the pre-depreciation period. This could be achieved by a combination of increasing the number of arrivals, the length of stay, and average daily expenditures. The magnitude of this task can be illustrated by considering how this might be achieved by increasing the number of arrivals. Taking the 7.6 million arrivals in 1996 as the base, arrivals in 1998 would have to reach 12 million to achieve the same revenues in foreign exchange as in 1996. The question is therefore whether this is feasible and, if so, what combination of policy options and private sector strategies could bring it about.

THE FEASIBILITY OF EXPANDING TOURISM IN THAILAND

As Table 7.5 shows, in the expansion associated with Visit Thailand Year 1987, Thailand's tourism arrivals rose from 2.4 million to 4.2 million in the period from 1985 to 1988, an increase of 73 per cent over three years.[12] In this section, we shall consider the reasons for this outcome and make a comparison with the present situation to determine whether such an outcome is again possible.

We shall draw in part on the work of Michael Enright *et al.* (1997) which focuses on competitiveness at the industry level and which

Table 7.5   Visitor Arrivals to ASEAN Countries, 1986–88

	1986	% Change	1987	% Change	1988	% Change
Indonesia	825,035	10.1	1,050,014	27.3	1,301,049	23.9
Malaysia	1,053,821	4.5	1,145,800	8.7	1,239,181	8.1
Philippines	781,517	1.1	794,700	1.7	1,043,114	31.3
Singapore	3,191,058	5.3	3,678,809	15.3	4,186,091	13.8
Thailand	2,818,092	15.6	3,482,958	23.6	4,230,737	21.5
Total	**8,669,523**	**8.4**	**10,152,281**	**17.1**	**12,000,172**	**18.2**

Source: Pacific Asia Travel Association (PATA), *1988 Annual Statistical Report*.

extends Porter's (1990) earlier ideas. We shall also include concepts drawn from the work of Susan Strange (Strange, 1988; Stopford and Strange, 1991) and particularly the notion of structures within the international political economy (IPE). Enright first considers the inputs into the production process of labour, capital, and technology together with the degree of competition and the strategies of firms. In addition to the impact of demand, he also takes account of 'agendas and institutions' which can operate at national or international levels. This takes the model into the realm of political economy, making explicit the role of non-market actors including governments, and this offers a link with Strange's ideas of a state/market balance in the IPE structures, especially of production and finance. In the following analysis, we shall draw on these models to consider the principal causes of the changes of the late 1980s and to judge whether similar changes may perhaps occur in the late 1990s.

The 1980s: International Structures, Domestic Agendas, and Demand

The second oil crisis of 1979/80 led to a sharply rising Thai current account deficit that was exacerbated by a rising exchange rate. The US dollar rose on the exchanges, following the dramatic rise in US interest rates in 1980, as too did the baht which had been linked to the dollar since 1963. As a result, the current account deficit had risen to around 8 per cent of GDP by 1983. Thai policy, however, was predicated on the export of agricultural products and import substitution for manufactures. As Phongpaichit and Baker (1995: 148) comment, by 1984 'the government sought relief by promoting the export of services'. The export of labour to the Middle East and tourism were the principal sectors. The Board of Investment offered investment incentives for the construction of hotels and the State tourism authority's budget was increased. In 1985, the Thai Cabinet approved a State-led campaign that would promote 1987 as a special year to visit Thailand. The marketing angle was the King's sixtieth birthday which, being especially auspicious in Buddhist thinking, would be accompanied by wide ranging ceremonial events.[13] Whilst this was still in the planning stages, though, the international finance structure had a major impact on tourism demand.

The rise in the value of the US dollar forced the Thai Government to devalue the baht by 15 per cent in 1984. However, as is graphically illustrated in Figure 7.1, this was both overshadowed and magnified by the fall in the dollar to which the baht remained closely linked.

Figure 7.1   Movements in the US Dollar Exchange Rate, 1970–95

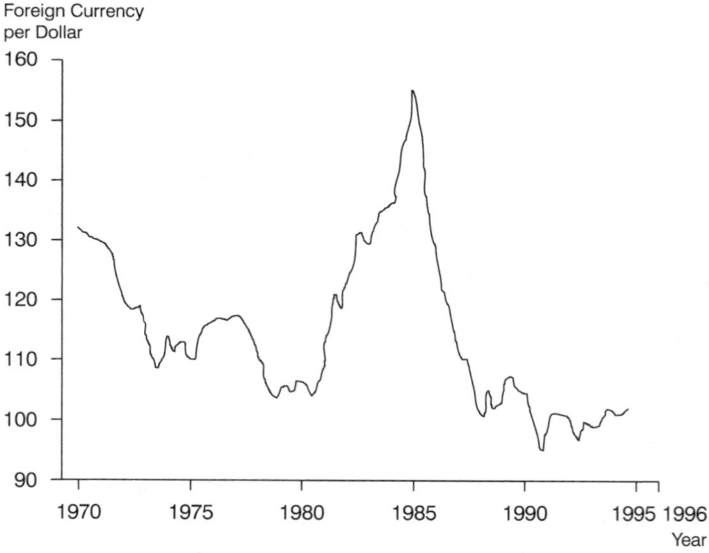

Source: Czinkota *et al.* (1996), 'International Business', Dryden Press.

Table 7.6   European Visitor Arrivals to ASEAN Countries, 1986–88

	1986	% Change	1987	% Change	1988	% Change
Indonesia	234,999	18.2	367,272	56.3	321,261	−12.5
Malaysia	158,026	5.3	165,400	4.7	132,077	−20.1
Philippines	91,084	2.5	91,240	0.2	110,375	21.0
Singapore	517,032	11.0	615,732	19.1	805,525	30.8
Thailand	550,731	21.0	719,402	30.6	904,665	25.8
Total	**1,551,872**	**14.2**	**1,959,046**	**26.2**	**2,273,903**	**16.1**

Source: Pacific Asia Travel Association (PATA), *1988 Annual Statistical Report*.

In early 1985, the dollar turned on the foreign exchange markets and, following the Plaza Agreement, plummeted. The yen soared, doubling in value against the baht by 1989.

Whilst the yen was the most affected currency, the German mark, sterling, and other European currencies also rose steeply. Table 7.5 shows the total arrivals to ASEAN countries in the three years to 1988, from which it can be seen that Thailand's growth rate was consistently above the average for the ASEAN countries as a whole. Only Indonesia's growth in arrivals actually exceeded that of Thailand, although the additional number of arrivals was far smaller. Table 7.6 shows the arrivals from Europe to ASEAN over the same period, and again the growth rate in Thailand surpasses the average rate for the ASEAN countries. Furthermore, the growth rate in 1986 was higher than for any other ASEAN country, which suggests that the currency effect was important and perhaps even more significant than the marketing campaign that followed.

The 1980s: Inputs, Competition, and Strategies

The core of the tourism product at the destination is hotel accommodation, which provides the key economic input. In Thailand, restraints on the construction of new hotels have been negligible and supply has been characterized by step jumps in the number of available rooms. One of the features characterizing 'Visit Thailand Year 1987' was the very high number of new hotels that had opened both in

Table 7.7   Tourist Accommodation in Thailand, 1981–86

		1981	1986
Total[1]	Establishments	436	2,669
	Rooms	32,173	116,997
Bangkok	Establishments	65	97
	Rooms	13,824	22,576

Note: (1) The totals are not strictly comparable because of differences in data collection in 1981 and 1986.
Source: Tourism Authority of Thailand, *Annual Statistical Report* (various issues).

Bangkok, still then the main destination, and in the emerging beach resorts of Pattaya and Phuket. Between 1978, when investors committed to new hotel projects, and 1986 the number of hotel rooms more than doubled nationwide. Table 7.7 gives an indication of the increases.

As new supply became operational from the early 1980s, prices began to fall. Hoteliers were faced with pressure in three areas. First, there was the increase in competition as new hotels opened. Second, the cost structure of hotels provides an incentive to discounting. Hotels function with relatively low variable costs but high fixed costs. New hotels face especially high fixed costs due to loan repayments, a feature that was particularly problematic then in Thailand. As US interest rates rose to a high of 18 per cent, Thai rates mirrored the rise as a result of the baht link to the dollar. Third, there was the negotiating power of foreign wholesalers and local tour operators, who used their intermediary role between the suppliers and the market to demand extremely advantageous hotel rates. Partly by default, and partly because of a lack of experience in the new hotels, the strategy of the Thai hotel sector from the mid-1980s was based almost exclusively on price discounting. The discounts were, naturally, offered on rates quoted in baht. As the baht was falling with the dollar from 1985, such discounts then further reduced the overall price, which was collapsing when measured in non-dollar linked currencies. The impact was significant. Between 1985 and 1988 arrivals increased at an average annual rate of 25 per cent.

It is clearly tempting for policy makers, strategists, and lobby groups to look back on the growth years of 1986 to 1988, to ascribe the growth largely to the 'Visit Thailand Year 1987' campaign, and to suggest that Thailand can do it again. The preceding analysis has argued, however, that the major cause of the expansion was price. Given this conclusion, it would appear that a further expansion of tourism arrivals might take place as a result of the currency depreciation in the late 1990s, which of course has implications for policy and strategy. However, before proceeding to a conclusion for the 1990s, we should make a slightly more detailed comparison between the two decades.

The 1990s: International Structures, Domestic Agendas, and Demand

The first direct comparison is with the external value of the Thai currency. The gyrations of the baht that accompanied the dollar's rise and fall in the mid-1980s caused relatively little stir within Thailand since the prime focus was on the stability of the baht against the dollar. This is in massive contrast to the concerns generated as the baht fell in 1997 against all currencies, apart from those even more affected by the Asian crisis. This then raises one key difference. The short-haul markets of Asia are unlikely to offer growth in tourism arrivals. Preliminary data for 1997 suggest that previous growth markets in Asia were depressed, with the possible exception of Hong Kong and Singapore. Overall growth in East Asian arrivals was a little over 1 per cent in 1997, with a decrease in the second half of the year. The softening of the yen and the problems in the domestic economy in Japan will render a repeat of the previous massive growth in Japanese arrivals virtually impossible. The North American market share has declined from around 6 per cent in the previous decade to 5 per cent in 1997, and hence remains relatively insignificant.

The one area, therefore, which has the potential for increased demand from a strong base is Europe. Hence, if a major contribution is to be made to Thailand's balance of payments, we may now ask whether the European market can provide that contribution. Whilst arrivals from Europe declined slightly in 1997, there was growth of 20 per cent in January 1998 year-on-year. The earlier decline can be explained by hedging strategies undertaken by the European package holiday companies who bought baht in the forward exchange markets before the depreciation. As a result, these firms were locked into the pre-depreciation exchange rate and hence were unable to reduce their selling prices.[14] Once the forward exchange contracts had been completed, however, package prices could fall to reflect the new exchange rate.

Finally, compared to the 1980s, manufactured exports had become far more significant and tourism less critical to the balance of trade. This argued for a lower priority on the government's agenda. Indeed, the tourism authority's budget was cut, in part because of the fiscal tightening mandated by the International Monetary Fund, and there is only a remote chance that this will be reversed without a major political initiative. Whilst tourism may offer a useful rhetoric in times of economic difficulty, it is unlikely to rise in importance. Ten years of export-led growth based on manufactures have changed the political priorities.

The 1990s: Inputs, Competition, and Strategies

The initiative for promotion will, therefore, have to rest with the private sector. A parallel with the 1980s has been a major leap in the number of available hotel rooms, with an attendant increase in competition. This increase in accommodation originated with the tourism boom associated with 'Visit Thailand Year 1987' and the continued expansion that lasted until the Gulf War. This provided the motivation for a very large number of new entrants into the hotel industry. Table 7.8 gives an indication of this increase. Businessmen who had succeeded in other industries latched on to this sector for three reasons. First, it appeared to make economic sense to invest where demand was increasing. Second, the real estate market was a prime beneficiary of the Thai boom, with much of the foreign corporate debt being channelled into speculative investments in condominiums and housing estates, office and retail complexes, golf courses, and, of course, hotels. Third, new entrants into the hotel sector were attracted by what one practitioner has termed the quest for 'return on ego, not return on investment':[15] a combination of status-seeking and imitative behaviour.

However, in common with much of Thailand's economic expansion during the 1990s, the increase in hotel accommodation was financed

Table 7.8   Tourist Accommodation in Thailand, 1988–96

		1988	*1996*
Total	Establishments	3,251	4,738
	Rooms	135,720	265,542
Bangkok	Establishments	116	186
	Rooms	25,605	45,569

Source: Tourism Authority of Thailand, *Annual Statistical Report* (various issues).

very largely by foreign loans. Low dollar interest rates, compared to rates available in the local currency, together with the conviction that the currency link was inviolate, left many investors with a serious foreign exchange exposure. From July 1997, therefore, debt service became a major cost. This applied not only to those entering the hotel business for the first time, but also to many established operators who raised additional funds for new projects and whose profits on existing operations were now committed for the additional debt service. As a result, there were no funds available in the private sector either for a new marketing drive. Without a sustained promotional strategy that could differentiate Thailand's tourism product, the hotels were forced back to a reliance on price cuts and on the marketing efforts of the tour operators and wholesalers.

CONCLUSIONS: CAN TOURISM HELP?

We may now return to this question, although from the above analysis the answer appears fairly clear and fairly negative. On the surface there are two similarities between the two decades. One is the effective devaluation of the baht against the currencies of Thailand's major tourism markets, and the other is the major increase in capital inputs in the form of new hotels. The recent effect of these has been to slash the price of the locally-produced part of the tourism product, in a similar manner to 1985. However, there are also significant differences between the circumstances in the two decades. In the first place, the short-haul markets, which given their proximity would normally have the greatest potential for growth, are the most troubled by the Asian economic crisis. Whilst Japan was not immediately affected by the events of 1997, the domestic recession and currency fluctuations do not suggest that this will be a growth market for some time and contraction of outbound tourism is more than likely in the immediate future. In the 1980s the Asian markets, and especially Japan, were responsible for a large share of the growth. Amongst the long-haul markets, North America has too small a share, and hence this leaves Europe as the only significant market with growth potential.

Between 1985 and 1988 European arrivals increased by an average of 33 per cent per year. If this were to be repeated, thanks largely to the price effect, then between 1997 and 2000 the number of European arrivals would rise to over 3 million. However, European growth alone would still only increase the total number of arrivals to 8.9 million by the new Millennium, up from 7.3 million in 1997. This is not impossible, but will not be sufficient on its own to have a significant economic effect on the external account.

We return to the issue of the foreign exchange earnings from tourism. Earlier, we argued that a 60 per cent increase in tourism arrivals would be needed simply to generate the same foreign exchange receipts in 1998 as in 1996, assuming that length of stay and average expenditures remained constant. In other words, 12 million arrivals would be needed in 1998 to maintain the status quo. The conclusion, therefore, is that tourism exports from ASEAN countries such as Thailand will not provide relief from the economic crisis. The only market in which significant growth could occur at present is Europe, but Europe alone is most unlikely to provide sufficient growth. Public policy measures are constrained by fiscal stringency that makes it virtually impossible for action to be implemented to support and promote tourism. In the private sector the cost structure of the hotels, and intense competition in an oversupplied hotel sector, are forcing down prices and hence reducing revenues in baht terms. In addition, the natural high fixed costs of hotel operations together with high levels of foreign debt service make it impossible to undertake any new marketing strategies.

The lessons from this study are that volumes will increase as a result of the present round of price reductions. Unlike the previous experience which occurred during a period of Asian economic expansion, this will happen during an economic downturn in Asia and growth will most likely come from Europe, as budget tourists are able to afford long-haul Asian holidays. Growth will not be sufficient, however, to compensate for the fall in the value of the currency and hence foreign currency earnings will actually fall. If tourism is to become, once again, a tool for the relief of current account difficulties, then the only solution would be to search for other markets that could provide the massive additional volume required to expand revenues as well as volumes. Perhaps the final lesson is that the focus should turn away from traditional tourism markets, and that policy makers and firm strategists should plan to diversify and to turn their attention to the non-traditional markets such as China. Whilst, at present, regulations on outbound travel act as constraints to the growth of tourism from the People's Republic, changes are possible. As a final conclusion, therefore, it would appear from the foregoing analysis that, in the short term, it is unlikely that tourism can provide the export-led recovery that industry lobby groups perceive is possible. However the current crisis should be a stimulus to longer-term thinking so that, when new markets emerge, destinations such as ASEAN are ready to take advantage of their potential.

NOTES

1. *Tourism Highlights 1996* (World Tourism Organization: Madrid).
2. Tourist expenditures are a frequently cited measure, but care must be exercised when using these data as there is considerable room for inaccuracy in the sample surveys generally adopted for their estimation.
3. *Tourism Highlights 1996* (World Tourism Organization: Madrid).
4. Brunei, and the new smaller members of ASEAN, are not included as tourism presently plays only a minor role in their economies, and tourism data are not reported to the Pacific Asia Travel Association (PATA). However it should be noted that in Brunei developments are underway, spearheaded by the Sultan, to develop tourist attractions in the very near future.
5. Industry sources in Bangkok.
6. *Bangkok Post*, 11 April 1998, Business Section.
7. WTTC (1998) 'Asia–Pacific Travel and Tourism in the New Millennium: a WTTC vision'.
8. WTTC press release: 6 March 1998, Bangkok.
9. It has long been a contention amongst proponents of tourism development that there is a significant multiplier effect that is gained from the

injection of international tourism revenues into national income flows. This concept underpins the 'satellite accounting system' now being proposed by the WTTC not only for Thailand but also for adoption as a global standard for the measurement of both the direct and indirect economic impacts of tourism. The result, according to the WTTC, is that the total economic value of goods and services attributable to tourism worldwide in 1996 is US\$3,600 billion (cited in *The Economist*, 10 January 1998).

10. Industry sources in Bangkok.
11. *Financial Times*, 3 April 1998.
12. In 1990 Malaysia imitated the Thai marketing theme, proclaiming it as 'Visit Malaysia Year'. Unfortunately statistical problems make it difficult to judge the outcome as arrivals data for Malaysia in 1989 and 1990 were compiled on a different basis (*PATA Annual Statistical Report*, 1989 and 1991).
13. This represents the completion of the fifth 'cycle' of life, each cycle being measured by the twelve years of the Asian calendar.
14. Industry sources in Bangkok.
15. Interview with Bert van Welbeck, Bangkok, April 1998.

BIBLIOGRAPHY

Czinkota, M. *et al*. (1996) *International Business* (Dryden Press).
Enright, M.J., Scott, E.E. and Dodwell, D. (1997) *The Hong Kong Advantage* (Hong Kong: Oxford University Press).
Hitchcock, M., King, V.T. and Parnwell, M.J.G. (1993) 'Tourism in South-East Asia: Introduction'. In M. Hitchcock, V.T. King and M.J.G. Parnwell (eds), *Tourism in South-East Asia* (London: Routledge).
Phongpaichit, P. and Baker, C. (1995) *Thailand: Economy and Politics* (Kuala Lumpur: Oxford University Press).
Porter, M.E. (1990) *The Competitive Advantage of Nations* (New York: The Free Press).
Sinclair, M.T. and Vokes, R.W.A. (1993) 'The Economics of Tourism in Asia and the Pacific'. In M. Hitchcock, V.T. King and M.J.G. Parnwell (eds), *Tourism in South-East Asia* (London: Routledge).
Stopford, J. and Strange, S. (1991) *Rival States, Rival Firms* (Cambridge: Cambridge University Press).
Strange, S. (1988) *States and Markets* (London: Pinter Publishers).

8 The Effects of Outward Direct Investment on the Performance of Locally-Controlled Companies in Malaysia

Jim Slater and Isabel Tirado Angel

INTRODUCTION

One of the most significant developments in international trade and investment over the past decade has been the growth of investment flows by firms from developing countries, such as Hong Kong, Singapore, South Korea, and Taiwan. Investment flows from developing countries have increased from 3 per cent[1] of total world outflows for the period 1970–90 (United Nations, 1992: 291), to 15 per cent[2] in 1995 (UNCTAD, 1997: 5); a trend which seems likely to continue. Furthermore, recent economic studies have shown that the real value of outward direct investment (ODI) from developing countries has increased at a higher rate than inward direct investment (Tolentino, 1987). Although the percentage of ODI from developing countries is small, most (89 per cent in 1996) originated from a small number of developing Asian nations (nine in 1996), which suggests that the flows are of some importance to these economies (United Nations, 1997: 149). These nine countries included the four Asian Newly Industrializing Economies (NIEs), four developing countries in Southeast Asia, and China (United Nations, 1997: 83).

The Malaysian Government has, for many years now, relied on foreign investors for the acquisition of advanced technology and relevant skills to enable further economic growth, and has provided generous government incentives to encourage inward foreign direct investment (FDI). However, recent studies have shown that the

179

transfer of technology and skills to local industry has been a rather slow and limited process (Astbury, 1996: 58; Nesadurai, 1994: 303; Rasiah, 1992), not in line with the national development targets which envisage a *fully developed* economy by the year 2020.[3] Consequently, local companies have been encouraged to hasten the learning process by investing overseas and sourcing technology from abroad. For example, Premier Choice is a company established at the beginning of 1996 by 20 leading Malaysian industrialists and financiers, to source technology from abroad through the purchase of stakes in high-technology Japanese, North American, and European companies whose technology can then be imported into Malaysia (Astbury, 1996: 60). Similarly, the Lion Group has purchased tyre technology from the Avon Group in Britain, and is now exporting to 30 countries. High levels of technology cooperation grants, acting as subsidies for the investors, are provided to induce overseas investment from this technology-hungry country.

The Malaysian Government was among the first of the ASEAN developing countries to actively promote overseas investment by their companies. The detail of the policies is similar to those of Singapore – see Chapter 4 for details. The main objectives of the incentives were to broaden the earning capacity of the country and to reduce the large factor payments abroad in the Balance of Payments – see Table 8.1. The Ministry of International Trade and Industry (MITI) set out three basic objectives of the outward investment policy. These were: first, to develop new markets or access to third markets; second, to boost sales of local components and resources; and third, to utilize ownership advantages possessed in plantation, mining, and manufacturing. Labour bottlenecks were possibly an additional factor encouraging the government to promote ODI, along with the desire to expand markets beyond the relatively small scale of the domestic economy. Although the objectives behind government incentives were clear, questions remain as to the real effects of ODI on the performance of the investing firms.

This chapter provides a preliminary analysis of the effects of ODI on the performance of locally-controlled companies in Malaysia. The first section sets out the regulatory background regarding ODI. The following section details the sources of data on Malaysian ODI. The theoretical section sets out the *a priori* expectations and

Table 8.1   The Malaysian Balance of Payments, 1985–95
(US$m)

Year	Outward Direct Investment (annual flow)	Inward Direct Investment (annual flow)	Factor Payments Abroad
1985	210	695	2218
1986	249	489	1852
1987	209	423	1963
1988	206	719	1941
1989	282	1668	2179
1990	532	2332	1872
1991	389	3998	2473
1992	514	5183	3143
1993	1325	5006	3207
1994	1817	4342	3594
1995	2575	4132	4236

Sources: ODI data: Bank Negara, Malaysia. Inward Direct Investment and Factor Payments Abroad: IMF, *International Financial Statistics*.

discusses some of the conceptual and measurement difficulties. This is followed by empirical analysis both of ODI trends and using a simple regression model to predict performance. The final section summarizes and suggests avenues for further research.

THE REGULATORY BACKGROUND

As ODI occurs, the role of the government shifts to ensuring that the flows of investment do not starve the economy of necessary capital, and that the benefits of the investment spill over into the economy as a whole. This balancing act is not easy, particularly for Malaysia for which exports have been necessary to sustain growth in the light of a small domestic market in terms of both population and purchasing power.

Malaysia has used both fiscal incentives, in the form of tax write-offs, and institutional incentives to promote ODI. Prior to 1991, the government's attitude towards ODI was that it should neither be punished nor encouraged. Income derived from overseas and remitted to Malaysia was subject to income tax and allowed tax credit (Yap, 1996: 4). The 1991 incentives were in the form of tax abatements on income earned overseas and remitted back to Malaysia, and tax reductions for pre-operating expenses (e.g. the cost of market research) (MITI, 1995: 252). In the main, these incentives were aimed at encouraging the repatriation of income by investors creating further benefits for domestic operations. In short, the benefits from overseas investment were expected to be realized in the medium-term, once the additional profits brought back by way of repatriation had been reinvested.

Further incentives were introduced in 1995, in the form of full tax exemption for repatriated earnings. In tandem, following the example of Japan and South Korea which had successfully exploited this strategy to gain rapid access to American electronics technology (Wong, 1986: 22), Malaysia and Singapore signed an agreement in November 1995 establishing a Third Country Investment Facility Fund, aiming for enterprises jointly to explore opportunities in third countries. Fifty per cent of initial expenditure would be reimbursed (up to US$80,000).

These incentives and promotional activities seem to have paid off, as Malaysia had one of the largest flows of funds from the ASEAN developing countries – see Chapter 4 for further details – and a substantial overseas stock close to US$9 bn in 1995. The agriculture sector has shown a sharp recovery in overseas investment since 1990, of which investment in other companies (i.e. not through subsidiaries or branches) has been a fundamental part. The tin mining industry as a whole started rationalizing its investment abroad in 1990, but a sharp increase in ODI from local companies has taken place since 1990. Excluding the banking sector, manufacturing is the main investor overseas. This sector has, since 1985, had a substantial share of other foreign assets, and direct investment increased so as to constitute 34 per cent of all foreign assets by 1994, and has continued to increase up to the present. It is important to bear in mind that this picture of the manufacturing sector relates only to *locally-controlled companies*. However, 70.7 per cent of the total 1994 output of the manufacturing sector was produced by foreign-controlled firms (Department of Statistics, 1994: xxv) and the analysis undertaken here may not hold for foreign-controlled companies.

DATA SOURCES ON MALAYSIAN OUTWARD DIRECT INVESTMENT

Malaysian ODI data are collected in several ways. On the one hand, the Central Bank (Bank Negara) collects data on ODI flows through the Cash Balance Reporting System.[4] On the other hand, the Department of Statistics (DOS) collects data on ODI flows and stocks through the annual Financial Survey of Limited Companies.

The figures from the two sources are not directly comparable – see Table 8.2 – for three main reasons. First, Bank Negara collects data on all international cash transactions by local enterprises,[5] both locally-controlled and foreign-controlled, whilst the DOS survey only covers a sample of the limited companies registered in Malaysia. However, the sample is quite substantial[6] and, in 1994, it covered 83 per cent

Table 8.2 Malaysian Outward Direct Investment, 1985–95 (US$m)

Year	Data from Bank Negara		Data from Dept of Statistics[1]	
	Flows	*Stocks*	*Flows*	*Stocks*
1985	210	749	−311	5984
1986	249	770	950	6736
1987	209	1023	−33	6910
1988	206	1469	1402	8415
1989	282	1751	−808	8161
1990	532	2283	760	8917
1991	389	2672	−322	6427
1992	514	3186	−262	6743
1993	1325	4511	1118	7976
1994	1817	6328	2763	10977
1995	2575	8903	—	—

Notes: (1) The DOS data include equity investment (though not reinvested earnings) and long-term loans, together with disinvestments, and repatriated capital. The DOS flows data include investments, long-term loans, and changes in other foreign assets. The DOS stock data include all foreign assets, except foreign securities.

Sources: UNCTAD (1997) for the Bank Negara data; Department of Statistics, Malaysia, *Report of the Financial Survey of Limited Companies*.

of the total revenues of companies in Malaysia (Department of Statistics, 1994: xi). Second, the DOS data include equity investment (though not reinvested earnings) and long-term loans, together with disinvestments[7] and repatriated capital. Hence, there are several years which show negative outward investment. Third, the Bank Negara data include official assistance to other countries in the form of long-term capital,[8] whereas such flows are not recorded by the DOS survey.

The major advantage of the DOS survey,[9] and the reason why the data from the survey are used in this chapter, is that the companies are therein classified as to whether they are locally-controlled (if 50 per cent or less of the shares are owned by foreigners) or whether they are foreign-controlled (if more than 50 per cent of the shares are owned by foreigners). In the 1994 survey, 69 per cent of the total revenues reported came from locally-controlled companies – see Table 8.3 – representing 57 per cent of the 1994 revenue of all Malaysian companies (Department of Statistics, 1994: xxiii–xxv).

Table 8.3   The Importance of Locally-Controlled Companies in Malaysia, 1985–94

Year	Numbers of Companies in Financial Survey[2]		Total Assets of Companies[1] (US$m)		Total Revenue of Companies (US$m)	
	Total	of which locally-controlled	Total	of which locally-controlled	Total	of which locally-controlled
1985	2897	2513	5984	4166	52240	36774
1986	2770	2320	6736	6232	44194	31587
1987	2772	2302	6910	6199	48184	33527
1988	2982	2427	8415	7376	55875	38650
1989	3237	2612	8161	6290	66954	45831
1990	4262	3392	8917	7597	89841	63048
1991	4672	3685	6427	4989	105159	72851
1992	5510	4384	6743	5348	127734	87877
1993	5942	4703	7976	6449	147813	102471
1994	6322	4986	10978	9174	177014	122066

Notes: (1) 'Total assets' include all foreign assets except foreign securities.
(2) The survey only includes companies with turnover of more than RM 5 m.

Source: Department of Statistics, Malaysia, *Report of the Financial Survey of Limited Companies*.

Typically, large proportions of the ODI from developing countries are attributable to foreign-controlled companies or the branches and/or subsidiaries of multinationals (Low *et al.*, 1996). In Malaysia, the figures are relatively modest (14.1 per cent of total investment, and 20.8 per cent of net reported profits in 1994) but nevertheless it is important to be able to focus on the activities of locally-controlled companies alone.

THEORETICAL CONSIDERATIONS

The link between investment and performance is not clear-cut. There is plenty of evidence at both firm and industry level that internal funds have a positive and significant effect on investment spending. However, the empirical literature on the effects of investment spending on earnings or cash flow is not very extensive. Kalecki was one of the earliest authors to provide statistical evidence for the effect of past investment on profits, by means of regressing US annual data over 11 years between 1929 and 1940. The relation was found to be positive (Kalecki, 1971: 90–2). He further argued that investment occurs before profits, because investors can decide to invest more than in a preceding period, but earnings is not a decision variable (Kalecki, 1971: 78–9): the direction of causation runs from investment to profit. Later studies such as McFetridge (1978) and Shapiro *et al.* (1983) have confirmed these findings. More recently, Mahdavi *et al.* (1994) tested whether there was a significant relationship between the stock of investment and earnings, again broadly establishing investment as the driving force in the relationship. Few studies have attempted to link ODI with performance. One, relevant to Malaysia, is that by Mohamed (1993). He found a link between FDI and performance in Malaysia, but the dataset covered the period before the introduction of the 1991 incentive programme.

Nor is it clear theoretically what the relationship is between ODI and performance. On the one hand, ODI from developing economies may serve long-term, through exposure, to enhance performance through improvements in managerial efficiency and through technology transfer. On the other hand, the Product Life Cycle (PLC) hypothesis suggests that, as the domestic industry moves through the cycle, ODI will shift activities to countries at lower stages of development. The seminal work by Vernon (1966) argues that ODI should occur mainly in growing and mature industries. Wells (1966) identified capital-intensive and concentrated industries as the main sources of ODI. Audretsch and Woolf (1986) found the focus to be in the maturity stage. Grant (1995: 239) and many recent texts simply characterize ODI as increasing through the cycle from the growth phase through maturity and even into decline. The PLC literature is thus not explicit on the relationship between ODI and performance. In the introduction phase, companies are probably making a loss due to investment, despite the existence of high margins. During the growth phase, industry profits emerge along with rapid sales growth. Significant exports (and few imports) are present and companies are likely to incur massive expenditure. In the mature phase, profits are expected to peak, exports to fall, and significant imports to appear. The need for continuous investment may diminish. However, one might surmise that the expectation of diminishing profitability in home markets drives overseas expansion.

There are also a number of conceptual and measurement difficulties in establishing a statistical relationship between ODI and performance. First, there is the question of how to measure performance. Here performance is measured by the rate of return on capital,[10] and is calculated as the ratio of the annual earnings from investment overseas to the stock of investment at the end of the year. We also calculate the gross profit margin: the ratio of gross profits to total revenue. Second, there is the issue of the time lag between investment and the materialization of the earnings therefrom. Here, as indicated above, we implicitly assume that earnings flow from the entire stock of capital. Third, there is the problem that data are not available for individual enterprises but only for industrial sectors. However, empirical studies suggest that aggregate results do seem to be a reflection of firm-specific behaviour (Fazzari and Mott, 1986–87) whereby variations in aggregate investment are shown to reflect variations at the level of the firm (Mahdavi *et al.*, 1994: 490). Fourth, there may be reporting differences between firms which can give rise to difficulties in the interpretation of accounting data. However, the Department of Statistics explicitly addressed this problem in the preparation of their statistics, and has endeavoured to minimize any potential bias (Department of Statistics, 1995: xxix).

Finally there are two potential problems which cannot be addressed. The data used relate to locally-controlled companies only. A change of ownership (from foreign to local control, or vice versa) may generate year-on-year discontinuities in the data, which may be particularly significant at the sectoral level. Unfortunately, however, we have no way of identifying or of controlling for such instances. There is also no way of knowing whether the capital flows represent investment that will eventually create a wider Malaysian economy, or whether they are simply capital flight.

EMPIRICAL ANALYSIS

The analysis begins with an examination of trends in overseas investment, and the earnings received from this investment, over the period 1985–94. The DOS survey provides a breakdown of the components of both the 'total foreign assets' and the 'overseas earnings' of locally-controlled companies. The 'total foreign assets' include: (a) investment in or claims on branches, subsidiaries, and affiliated enterprises abroad; (b) investment in other companies abroad at book value; (c) long-term loans to non-residents; (d) holdings of foreign securities; and (e) other foreign assets. The 'overseas earnings' include: (a) overseas profits made up of undistributed profits (or losses) accruing from subsidiaries abroad, and dividends and interest accruing from branches or subsidiaries abroad; (b) other overseas dividends accruing from abroad; (c) other interest accruing from abroad; (d) dividends payable to non-residents; and (e) interest payable to non-residents. Table 8.4 highlights the contributions of outward direct investment in branches, subsidiaries etc. to both assets and earnings.

ODI through subsidiaries or branches represented only 32 per cent of the total foreign assets held in 1994 by locally-controlled companies in Malaysia: an average annual growth rate of 27 per cent since 1990. The earnings from this investment accounted for 9.7 per cent of total earnings, up 18 per cent per annum since 1990. When the foreign assets of branches are excluded – see Table 8.5 – ODI rises to 43 per cent of total foreign assets, and earnings to 53 per cent of the total.

Table 8.4 The Return on Outward Direct Investment by Locally-Controlled Malaysian Companies, 1994

Industry	Investment in Subsidiaries or Branches as % of Total Foreign Assets	Earnings from Investment in Subsidiaries or Branches as % of Total Overseas Earnings	Return on Capital Invested in Subsidiaries or Branches Overseas (%)	Return on Total Foreign Assets (%)
Rubber	10.4	98.6	46.9	9.9
Other Agriculture	15.7	88.1	6.3	4.7
Tin Mining	99.9	86.9	11.0	12.7
Other Mining	0.06	0.0	0.0	2.9
Manufacturing	34.5	21.9	1.5	2.7
Construction	13.9	0.0	0.0	2.1
Wholesale Trade	72.3	5.4	2.1	2.9
Retail Trade	99.9	0.0	0.0	8.7
Banks and Financial Institutions	31.9	6.4	8.9	5.4
Other Industries	37.0	27.0	1.5	2.4
All Industries	32.4	9.7	1.2	4.7

Source: Department of Statistics, Malaysia, *Report of the Financial Survey of Limited Companies.*

Table 8.5   The Stock of Malaysian Overseas Assets and Overseas Earnings, 1985–94

Year	Stock of Foreign Assets (at end year)		Overseas Earnings	
	Total (US$m)	Investment in Subsidiaries as % of Total Stock of Foreign Assets[1]	Total (US$m)	Earnings from Investment in Subsidiaries as % of Total Overseas Earnings
1985	5984	22.32		
1986	6736	20.98		
1987	6910	24.53		
1988	8415	19.80		
1989	8161	26.67		
1990	8917	22.72	562	61.99
1991	6427	38.37	507	36.24
1992	6743	36.83	378	50.10
1993	7976	39.93	491	48.95
1994	10977	43.39	536	53.33

Notes: (1) The foreign assets of branches are excluded.
Source: Department of Statistics, Malaysia, *Report of the Financial Survey of Limited Companies*.

Table 8.4 also presents estimates of the returns on both ODI and total foreign assets for each industrial sector. These returns are calculated as the overseas earnings for the year as a percentage of the stock of foreign assets at the end of the year. The overall return on ODI was 1.2 per cent in 1994.[11]

At the sectoral level, the highest return is in the rubber industry (47 per cent), followed by tin mining (11 per cent), banks and finance (8.9 per cent), and agriculture (6.3 per cent).[12] The table also shows that the overseas earnings of the primary industries such as rubber, agriculture, and tin mining depend almost entirely on earnings through subsidiaries and branches. In short, the more upstream the sector, the higher the proportion of direct investment earnings in total earnings. There are a number of possible explanations for this. It could be that greater Malaysian managerial control generated higher returns. Or it could be that ODI was undertaken for profit rather than strategic reasons. Or it could be that the methods of data reporting tend to inflate the return on ODI. The trade sectors (both wholesale and retail) displayed small earnings despite their heavy reliance on ODI. It is believed that the overseas earnings from licensing and franchising agreements, so popular in the trade sector, is classified as 'other interest accruing from abroad' rather than as ODI earnings, and hence the return in this sector is underestimated.

Ordinary Least Squares regression analysis was used to analyse further the relationship between ODI and performance. Data on first differences[13] were calculated for five industries[14] for the years between 1969 and 1994: i.e. all continuous variables were expressed as percentage changes over the previous year. Two alternative measures of the profit margin were used as the dependent variable: one the ratio of gross profits to sales; the other the ratio of gross profits to total revenues.[15] A number of explanatory variables were initially incorporated,[16] but many were later omitted because of a lack of statistical significance. The final results (using both alternative measures of the dependent variable) highlighted just three variables as being statistically significant – see Table 8.6. The first was the book value of ODI in branches and subsidiaries overseas. The second

Table 8.6　The Determinants of Profit Margins in Malaysian Industry: Regression Results

Explanatory Variables	Dependent Variable			
	gross profit sales	*gross profit sales*	*gross profit revenue*	*gross profit revenue*
Outward Direct Investment	-9.25 $p = 0.024$	-7.99 $p = 0.048$	-8.29 $p = 0.0139$	-7.24 $p = 0.0261$
Total Domestic Assets	-44.2 $p = 0.043$	-60.49 $p = 0.0057$	-37.34 $p = 0.0373$	-50.9 $p = 0.0046$
1991 ODI Incentives		45.38 $p = 0.0028$		33.85 $p = 0.0024$
Adjusted R^2	0.09	0.155	0.103	0.17
Durbin Watson	2.1548	2.3323	2.1194	2.2841
Significance of F-ratio	0.002	0.0001	0.0009	0.0001
Sample size	112	112	112	112

was total domestic assets. And the third was a dummy variable[17] introduced to capture the effect of the 1991 government incentives for ODI.

The adjusted coefficients of determination were low in all the regression equations, but this is to be expected when investment data are used (Baskin, 1989: 33) and when those data are expressed as first differences (Maddala, 1988: 191). Nevertheless, further investigation is required and the following conclusions should be regarded as extremely tentative. The negative coefficients for ODI are consistent with our expectation from the PLC hypothesis that performance will decline later in the life cycle as production is moved overseas. But of most interest are the positive coefficients for the dummy variable used to capture the effects of the 1991 incentives. The introduction of these incentives appears to have had a marked effect upon the profit margins of locally-controlled companies.

CONCLUDING REMARKS

We have endeavoured in this chapter to explore the relationship between ODI by locally-controlled companies in Malaysia, and their performance. Although there are considerable problems with the interpretation of the basic data and our results are tentative, it appears that the 1991 incentives introduced by the Malaysian Government did have a significant impact upon profit margins within the rubber, agriculture, mining, and manufacturing industries.

Why should this be? On the one hand, tax-free earnings may have been brought back to the country, possibly to be re-invested stimulating further local investment. This increase in the quantity and quality of the domestic physical capital may eventually translate into higher productivity rates, as well as increased competition and more efficient investment. On the other hand, an increase in actual overseas investment after 1991 may have been of the type that uses the industrial structure in Malaysia to its benefit, by buying in the domestic economy, and thus stimulating further local investment and creating a virtuous circle. The empirical evidence may also suggest the presence of additional benefits in the form of managerial and technical expertise acquired directly through overseas activities.

However, it is difficult to assess the exact mechanisms without more detailed data.

The results support the view that industries do react to external stimuli, in this case government incentives for overseas investment. Nevertheless, the increased outflow of funds for ODI has its dangers. Despite much progress over the years, the Malaysian economy is somewhat fragile with several sectors, such as textiles, petrochemicals, and electronics, still largely dependent on foreign investors. A widening of the domestic industrial base controlled by local companies would be desirable at the same time *as* overseas investment, to compensate for the initial development benefits forgone to the industries where ODI takes place. Eventually in time, with the help of government incentives, sufficient profits could be repatriated thus directly benefiting the industries involved in ODI.

We conclude by reiterating the tentative nature of the conclusions in this chapter, and drawing attention to possible avenues for further analysis. First, some industries are more cyclical than others and, if they operate with high levels of gearing, a sharp fall in profits will occur if demand falls (i.e. these industries have a higher volatility of earnings) (Ellis and Williams, 1993: 264). This should be taken into account in future analysis. Second, related industries may indirectly benefit more from overseas investment than the industries actually undertaking the investment, through subcontracting for example. This situation is clearly rather difficult to portray in this kind of analysis, and some kind of input–output framework would be required. Third, the low adjusted R^2 may suggest that investment flows should be incorporated in the analysis, along with variables portraying the determinants of market structure. Perhaps the correct method of analysis should involve a simultaneous framework allowing for the spillover effects of overseas and foreign investment to be captured.

NOTES

1. 3 per cent corresponds to US$35 bn.
2. 15 per cent corresponds to US$51 bn.
3. These targets are collectively called 'Vision 2020'.
4. Prior to 1991, data collection was through the Exchange Control Records, and data are available as far back as 1980.

5. Malaysian companies do not require approval for overseas investment if the capital outflow is below US$4 m per year (in 1996); above that amount or if a credit facility is in place for up to US$2m from the Malaysian financial sector, approval is required by Bank Negara. This means that 'small' investments are not recorded in the Bank Negara statistics, whereas they are picked up in the DOS survey.

6. The sample only covers larger enterprises (i.e. those with annual turnover in excess of 5 million ringgit) and should not be considered as representative of the population of companies though, as it is the larger companies who are more involved overseas, it does capture the major proportion of these overseas activities. The RM5 m cutoff applies to the surveys for 1979–94. Between 1973 and 1978, the cutoff was RM1m, and all companies were included between 1969 and 1972. There is thus some lack of comparability in coverage, but this is not thought to be significant as it was not until the 1980s that ODI became important.

7. ODI values change depending upon the classification principles used by different countries. The traditional asset/liability principle includes all intra-group assets as direct investment abroad, and intra-group liabilities as direct investment at home. The new IMF directional principle, however, suggests that direct investment at home should be treated independently from direct investment abroad: i.e. all flows are recorded on a directional basis, from the investor to the enterprise. Should a subsidiary take an interest in the original investor, this transaction is recorded as a disinvestment: i.e. the liabilities of the parent to the overseas subsidiary should be classified as a deduction in domestic investment abroad.

8. Such as to Pakistan, Iraq, and Myanmar under the Palm Oil Credit and Payments Agreement (Bank Negara, 1996: 287). Credit extended to Iraq and Myanmar in 1994 was offset by payments from Pakistan under a previous arrangement.

9. Enterprise surveys are also widely recognized as the most complete source of investment on investment (IMF, 1995: 153).

10. See Fisher and McGowan (1983) for a critique of the use of accounting rates of return as measures of profitability.

11. This figure had varied between 1.2 per cent and 1.7 per cent over the period 1990 to 1994.

12. Asian investors have been particularly interested in tropical agriculture and have, for instance, invested heavily in palm oil in Africa giving new dynamism to traditional sectors (World Bank, 1997).

13. Several modifications were made to the data. All differences were scaled downwards by a factor of 10, so the values of all the series lay in the range between -10 and $+10$. This procedure reduces heteroskedasticity. Occasionally when a measurement was thought to be an outlier, and further scaling would otherwise minimize the measurements for the other years, that outlier was allowed to have a value

outside the range. Augmented Dickey–Fuller tests were applied to detect non-stationarity in the series. All series became stationary at the 1 per cent level of significance after taking first differences. This procedure reduces the problem of serial correlation. It is important to bear in mind that causal relationships may be spurious due to non-stationarity in the raw series. Stationarity was tested using the F-statistics, under the null hypothesis of no causal relationship.

14. Rubber, other agriculture, tin mining, other mining, and manufacturing. Following Audretsch (1987), tin mining was found to be in decline from 1982, and so data for 1982–94 were excluded from the analysis. Only 12 observations were therefore recorded for tin mining.

15. Total revenues are equal to sales plus other revenues (e.g. interest).

16. Total revenues, sales, profits, expenditure on fixed assets, imports, exports, domestic assets, equity share capital, the book value of ODI in branches and subsidiaries overseas, the book value of investment in other companies overseas, long-term loans to non-residents, other foreign assets, net undistributed profits and dividends from investments in companies, other dividends received, and other interest received. See Ellis and Williams (1993: 37), Grant (1995: 231), Wheelen and Hunger (1995: 97), Lynch (1997: 125), and De Wit and Meyer (1994: 376).

17. This dummy variable had a value of unity for the years 1991–94 (the policy-on period) and a value of zero for the years 1969–90 (the policy-off period).

BIBLIOGRAPHY

Alexander, G.J. and Buchholz, R.A. (1978) 'Corporate Social Responsibility and Stock Market Performance', *Academy of Management Journal*, 21, 3, 479–86.

Arlow, P. and Gannon, M.J. (1982) 'Social Responsiveness, Corporate Structure, and Economic Performance', *Academy of Management Review*, 7, 2, 235–41.

Astbury, S. (1996) 'Technology Transfer: Struggling to Acquire Expertise', *Asia Business*, July, 58–62.

Audretsch, D.B. (1987) 'An Empirical Test of the Industry Life Cycle', *Weltwirtschaftliches Archiv*, 123, 297–307.

Audretsch, D.B. and Woolf, A.G. (1986) 'The Industry Life Cycle and the Concentration of Profits Relationship', *American Economist*, 30, 2, 46–51.

Bank Negara, Malaysia (1996) *Annual Report 1995* (Kuala Lumpur: Bank Negara).

Bar-Yosef, S., Callen, J.L. and Livant, J. (1987) 'Autoregressive Modelling of Earnings-Investment Causality', *Journal of Finance*, 42, 11–28.

Baskin, J. (1989) 'An Empirical Investigation of the Pecking Order Hypothesis', *Financial Management*, Spring.

Baumol, W.J., Heim, P., Malkiel, B.G. and Quandt, R.E. (1970) 'Earnings Retention, New Capital and the Growth of the Firm', *Review of Economics and Statistics*, 52, 345–55.

De Wit, B. and Meyer, R. (1994) *Strategy: Process, Content, Context. An International Perspective* (Minneapolis: West Publishing Company).

Department of Statistics, Malaysia (1969–95) *Report of the Financial Survey of Limited Companies*. 1969 to 1995 issues (Kuala Lumpur: Department of Statistics).

Downe, E.A. and Pan, W. (1992) 'Why does Business Invest?' *Journal of Post-Keynesian Economics*, 15, 1, 51–62.

Downe, E.A. and Pan, W. (1993–94) 'Why does Business Invest?: Rejoinder', *Journal of Post-Keynesian Economics*, 16, 2, 317–19.

Ellis, J. and Williams, D. (1993) *Corporate Strategy and Financial Analysis* (Glasgow: Pitman Publishing).

Fazzari, S.M. and Mott, T.L. (1986–87) 'The Investment Theories of Kalecki and Keynes: an Empirical Study of Firm Data, 1970–82', *Journal of Post-Keynesian Economics*, 9, 171–87.

Fisher, F. and McGowan, J. (1983) 'On the Misuse of Accounting Rates of Return to Infer Monopoly Profits', *American Economic Review*, 73.

Freeman, R.E. (1984) *Strategic Management: a Stakeholder Approach* (Boston MA: Pitman).

Grant, R.M. (1995) *Contemporary Strategy Analysis: Concepts, Techniques, Applications*. Second Edition (Oxford: Blackwell Business).

International Monetary Fund (1995) *Balance of Payments Compilation Guide* (Washington DC: IMF).

Kalecki, M. (1971) *Selected Essays on the Dynamics of the Capitalist Economy, 1933–1970* (Cambridge: Cambridge University Press).

Legal Research Board (1996) *Environmental Quality Act 1974 (Act 127) and Subsidiary Legislations (as of 1st November 1996)* (Kuala Lumpur: Legal Research Board).

Linter, J. and Glauber, R. (1972) 'Higgledy Piggledy Growth in America', In J. Lories and R. Brealey (eds), *Modern Developments in Investments Management* (New York: Praeger).

Little, I.M.D. (1962) 'Higgledy Piggledy Growth', *Bulletin of the Oxford University Institute of Statistics*, 24, 387–412.

Louton, D. and Domian, D.L. (1995) 'Dividends and Investment: Further Empirical Evidence', *Quarterly Journal of Business and Economics*, 34, 2, 53–64.

Low, L., Ramstetter, E.D. and Yeung, H.W-C. (1996) 'Accounting for Outward Direct Investment from Hong Kong and Singapore: Who Controls What?' NBER Working Paper no. 5588.

Lynch, R. (1997) *Corporate Strategy* (London: Pitman Publishing).

Maddala, G.S. (1988) *Introduction to Econometrics* (New York: Macmillan).

Mahdavi, S., Sohrabian, A. and Kholdy, S. (1994) 'Cointegration and Error Correction Models: the Temporal Causality between Investment and Corporate Cash Flow', *Journal of Post-Keynesian Economics*, 16, 3, 478–98.

McFetridge, D. (1978) 'The Efficiency Implications of Earnings Retentions', *Review of Economics and Statistics*, 60, 218–24.

McGuire, J.B., Schneeweiss, T. and Branch, B. (1990) Perceptions of Firm Quality: a Cause or Result of Firm Performance', *Journal of Management*, 16, 1, 167–80.

McGuire, J.B., Schneeweiss, T. and Sundgren, A. (1988) 'Corporate Social Responsibility and Firm Financial Performance', *Academy of Management Journal*, 31, 4, 854–72.

Ministry of International Trade and Industry, Malaysia (1995) *Malaysian International Trade and Industry Report 1995* (Kuala Lumpur: MITI).

Mohamed, R. (1993) 'The Market Structure of Selected Malaysian Manufacturing Industries', *Economic Bulletin for Asia and the Pacific*, 44, 1, 17–29.

Nesadurai, H.E.S. (1994) 'Public Sector R&D Institutes and Private Sector Industrial Technology Development'. In V. Kanapathy and I. Muhd Salleh (eds), *Malaysian Technology: Selected Issues and Policy Directions*, pp. 295–333 (Kuala Lumpur: ISIS).

Pava, M.L. and Krusz, J. (1995) *Corporate Social Responsibility: the Paradox of Social Costs* (Westport CT: Quorum Books).

Porter, M.E. (1980) *Competitive Strategy: Techniques for Analysing Industries and Competitors* (New York: The Free Press).

Rasiah, R. (1992) 'Foreign Manufacturing Investment in Malaysia', *Economic Bulletin for Asia and the Pacific*, 43, 1, 63–77.

Shapiro, D., Sims, W.S. and Hughes, G. (1983) 'The Efficiency Implications of Earnings Retentions: an Extension', *Review of Economics and Statistics*, 65, 327–31.

Strebel, P. (1992) *Breakpoints* (Cambridge MA: Harvard Business School Press).

Terry, E. (1996) 'An East Asian Paradigm?' *Atlantic Economic Journal*, 24, 3, 183–98.

Theilman, W. (1993–94) 'Comment on "Why does Business Invest? an analysis of industry accounting data" ', *Journal of Post-Keynesian Economics*, 16, 2, 313–15.

Tolentino, P. (1987) 'The Global Shift in International Production. The Growth of Multinational Enterprises from the Developing Countries: the Philippines'. Unpublished Ph.D. Thesis (University of Reading).

United Nations, Centre on Transnational Corporations (1992) *World Investment Directory 1992. Volume 1: Asia and the Pacific* (New York: United Nations).

United Nations, Economic and Social Committee for Asia and the Pacific (1997) *Economic Survey of Asia and the Pacific* (New York: United Nations).

United Nations Conference on Trade and Development (1997) *World Investment Report 1997: Transnational Corporations, Market Structure and Competition Policy* (New York: United Nations).

Vance, S.C. (1975) 'Are Socially Responsible Corporations Good Investment Risks?' *Management Review*, 64, 8, 18–24.

Vernon, R. (1966) 'International Investment and International Trade in the Product Cycle', *Quarterly Journal of Economics*, 80, 2, 190–207.

Vogt, S.C. (1994) 'The Role of Internal Financial Sources in Firm Financing and Investment Decisions', *Review of Financial Economics*, 4, 1, 1–24.

Waddock, S.A. and Graves, S.B. (1997) 'The Corporate Social Performance–Financial Performance Link', *Strategic Management Journal*, 18, 4, 303–19.

Wells, L.T. (1966) *The Product Cycle and International Trade* (Cambridge MA: Harvard University Press).

Wheelen, T.L. and Hunger, J.D. (1995) *Strategic Management and Business Policy*. Fifth Edition (Reading MA: Addison-Wesley).

Wong Poh Kam (1986) *Foreign Investment Obstacles and Opportunities* (Kuala Lumpur: ISIS).

World Bank (1997) 'Asian Investors see Improving Climate for Business in Africa'. Annual Meeting, Program of Seminars, September.

Yap Leng Kuen (1996) 'Malaysians Going Great Guns Cross-Border', *The Star* (1 January), 4–5.

9 The Distribution of Foreign Direct Investment in Vietnam: an Analysis of its Determinants

Bernadette Andréosso-O'Callaghan and John Joyce[1]

INTRODUCTION

A substantial body of literature has sought to identify and explain international flows of capital investment at the global level, whilst relatively less effort has been expended on the intra-national distribution of this investment. Moreover, the pioneering studies on the sub-national distribution of foreign direct investment (FDI) have concentrated on the case of Western economies (Glickman and Woodward (1988), McConnell (1990), and Ó hUallacháin and Reid (1992) for the United States; Hill and Munday (1992), Taylor (1993), and Collis and Noon (1993) for the United Kingdom; Thiran and Yamawaki (1995) for the case of Japanese Investment in the European Union; Andréosso-O'Callaghan (1996) for Japanese investment in France; Molteni (1997) for Japanese Investment in Italy). Much less attention has been conferred on the spatial distribution of FDI in the case of developing economies. In the case of Vietnam, and in spite of recent evidence of the uneven distribution of FDI within the country (IMF, 1996; US Embassy, 1996), research on FDI remains very much at the macroeconomic level.

Given the increasing openness of the various economies of the world, FDI is an important tool of regional development in many countries' economic strategies. Gaining knowledge on the intra-national distribution of FDI in a given country is thus paramount to the formulation of adequate regional policies aimed at lessening intra-national disparities (McDermott, 1977; Yannopoulos and Dunning, 1976). The contribution of transnational corporations (TNCs) to regional economic development spans from the simple backward and forward linkage effects highlighted in Hirschman's pioneering study (Hirschman, 1958),[2] to the fostering of technological and research linkages with research institutions, and to the building up of techno-logical clusters as discussed by Young *et al.* (1994).

The objective of this chapter is to highlight the explanatory power of the different variables that determine the spatial distribution of FDI within a country, namely Vietnam. The first section will describe the major FDI trends in Vietnam since the beginning of the eco-nomic renovation policy; the significance of EU foreign investment will be emphasized, and the spatial location of FDI will be high-lighted. The second section will provide a brief review of the possible factors determining the geographical distribution of FDI within a number of selected countries; another part of this section will encompass the econometric model, its results, and their interpreta-tion. Concluding avenues will finally be suggested.

ECONOMIC REFORM AND ITS IMPACT ON FOREIGN DIRECT INVESTMENT

Placed against a background of conflicts lasting nearly 40 years, the main motivation behind the economic reforms in the 1980s was less a preference for a definite switch-over to a capitalist regime than a threat of an imminent economic crisis resulting from economic isolation (Reidel, 1997; Kolko, 1997). Most of the economies of Southeast Asia had embarked upon a process of modernization which started at different points in time from the 1950s – see Chapter 2 for details. Taiwan and Hong Kong were among the first countries to initiate an import-substitution policy, which was subsequently replaced by an export-led approach. High growth rates were achieved through the stimulation of export-oriented industries, and the development of direct investment promotion policies. Nearly 40 years after the beginning of this new developmental approach, the need to catch up, and to avoid lagging behind others in the South-east Asian region prompted the economic renovation of Vietnam

(*Doi Moi*), a development model adopted at the Sixth Congress of the Communist Party in December 1986 (Schot, 1996; Tran Van Hoa, 1997; Ljunggern, 1997). Following the path already drawn by another centrally planned economy, the People's Republic of China, and showing the lead to other Indo-Chinese economies such as Laos and Cambodia, Vietnam committed itself to build socialism through the market by promoting equity and social justice for disadvantaged individuals and regions, and by preventing the alienation of urban workers to capitalist-style exploitation.[3] A balanced arrangement between State and market was envisaged within the reform framework. Vietnam's commitment to admission into the Asia–Pacific Economic Co-operation (APEC) forum, to the organization of the Sixth ASEAN Summit in Hanoi in December 1998, as well as to negotiating admission to the World Trade Organization, ensured continuous progress for the reforms and for the realization of targets laid down in the Sixth Five-Year Plan (1996–2000) (Ezaki and Son, 1997).

The accession of Vietnam to ASEAN in July 1995 was aimed at facilitating the reform process through the stimulation of trade and investment flows with its neighbour countries. Although a number of empirical studies have shown that the static gains arising from regional economic integration are small in the case of the ASEAN (Imada *et al.*, 1991), there was the prospect of substantial dynamic gains arising from an elimination of tariffs and non-tariff barriers in the emerging ASEAN Free Trade Area (AFTA), combined with the dynamic effects arising from economies of scale, specialization, increased competition, and increased foreign investment flows. Membership of ASEAN represents indeed an additional incentive for putative foreign investors who see Vietnam as offering scope for relocating manufacturing activities from higher-cost neighbour countries. TNCs can also use the country as a location for the manufacturing of intermediate products belonging to their vertically-integrated production processes spanning over the entire ASEAN region. Thus FDI promotes structural change in the host country and, as a result, economies of scale and industrial specialization set in, which in turn nurture intra-firm and intra-industry trade in the region as a whole. The increasing incidence of intra-industry trade gives rise to complementarities between the industrial structures of different countries in the region, which raises the degree of regional economic integration. Because of the high degree of complementarity existing between Vietnam and its neighbouring countries, the accession of Vietnam and of the other Indo-Chinese economies to ASEAN reinforces regional economic integration in Southeast Asia.[4]

Central to the reform process in Vietnam has been the Foreign Investment Law (FIL). Soon after the reunification of the country in 1975, the Vietnamese Government enacted a first law on foreign investment which, because of its many limitations and rigidities, failed to attract foreign investors. The revised law of 1987 eased many regulations, and the law was amended again in 1992. In November 1996, the National Assembly introduced further amendments to the FIL, which came into effect almost immediately.[5] The more relaxed regulations contained in the amended versions of the Law allowed for FDI to take place through the following forms: joint ventures, wholly foreign-owned enterprises, and contractual business cooperation between foreign and Vietnamese parties in which no corporate entity is formed (Schot, 1996). The Build-Operate-Transfer (BOT) company is often referred to as the fourth vehicle of FDI, even though it can take any one of the three forms of direct investment stipulated above. Although joint ventures are by far the preferred form of FDI, representing more than 70 per cent of the total investment capital approved (Apoteker, 1996), wholly-owned foreign enterprises are on the increase.

Preferential tax rates or incentives are available to investors who meet certain guidelines set out in the FIL and some decrees, such as the promotion of geographic remote areas and of exports and technology transfer. For example, whereas all forms of foreign-invested enterprises are subject to a standard profit tax rate of 25 per cent, lower rates apply to projects meeting at least two of a number of specified conditions, such as the use of advanced technology or meeting a minimum export requirement (20 per cent), location in mountainous regions or BOT projects (10 per cent) (Schot, 1996).

The ambition of the country is to become a modernized economy by the year 2020, through the implementation of an industrialization-by-invitation policy which allows access to foreign exchange, advanced technology, technical and managerial know-how (*Economic Development Review*, 1996). These measures have contributed to significant growth in Vietnam, and have changed a starvation- and inflation-laden country into an export-oriented, rapidly growing economy with rising incomes, and with highly motivated and skilled workers. GDP growth rates have been above or close to 8 per cent per annum since 1992 (EUROSTAT, 1998), and exceeded 9 per cent during the mid-1990s up until the onset of the Asian currency crisis.[6] FDI has helped the adjustment of the economic structure from an agrarian to an industrialized base, through the development of new industries and the creation of new jobs.

TRENDS IN FOREIGN DIRECT INVESTMENT IN VIETNAM

When assessed in a comparative framework, and bearing in mind its delayed opening to foreign investors and its lower level of economic development, Vietnam's process of capital accumulation clearly relies very heavily on FDI. This phenomenon is even more pronounced than in most other developing countries in both Asia or Latin America. The share of FDI in total gross domestic investment in Vietnam was above 22 per cent during the period 1991–95. This compares with less than 5 per cent for Thailand and Indonesia during the period 1990–94, and with approximately 10 per cent for the People's Republic of China[7] (Apoteker, 1996).

The Major Source Countries

Our analysis will focus on the evolving situation recorded since 1988, as hardly any foreign investment took place before the FIL came into force. Table 9.1 provides a breakdown of foreign investment in Vietnam by country of origin during the decade between 1988 and 1998. In line with the situation prevailing in other Asian countries, Vietnam's Asian neighbours dominate. The Asian newly industrialized economies (NIEs) secured more than one-half, and the ASEAN countries more than one-fifth, of total foreign investment in Vietnam over the period 1988–98. Singapore leads the way, with 16.9 per cent of total registered foreign capital over the period, followed by Taiwan (13.8 per cent), Japan (9.8 per cent), Hong Kong (9.5 per cent), and South Korea (8.9 per cent). Japan's share, which was initially very small, has been rising rapidly since the lifting of the US embargo in February 1994. In the same way, US investment has risen

Table 9.1   Foreign Direct Investment Stock in Vietnam at
end October 1998, by Country of Origin[1]

Country	Number of Projects	Total Capital[2]
Singapore	212	5500
Taiwan	453	4500
Japan	292	3200
Hong Kong	291	3100
South Korea	240	2900
France	136	1700
British Virgin Islands	69	1700
USA	91	1200
Thailand	111	989
Australia	80	940
Malaysia	72	926
United Kingdom	31	728
Panama	9	675
Netherlands	34	636
Switzerland	26	619
Others	n.a.	3287

Notes: (1) Cumulative FDI over the period from 1 January
1988 to 31 October 1998. The figures relate to
committed investment, or to investment for
which licences have been approved and officially
registered. Only a third of the committed
investment has been effectively disbursed.
(2) The capital is measured in US$ million.
Source: Ministry of Planning and Investment, Vietnam.

since 1994, although to a much lesser degree than Japanese investment. With 3.7 per cent of the total, US firms are the eighth largest
group of investors in Vietnam. France and Switzerland combine to
account for approximately 7 per cent of approved investment in
Vietnam over the period 1988–98. Apoteker (1996) endeavoured to
estimate the bias in the statistics created by the existence of offshore
investment facilities used for tax minimization purposes, such as the
British Virgin Islands and also Hong Kong. According to these estimates, around 12 per cent of total EU investment goes through these
'other' locations. Given that the average UK project in Vietnam is
nearly twice as big as the average French project, it is likely that the

share of UK firms in Vietnam's total FDI is rather higher than the low and official 2.2 per cent.

Japanese firms have concentrated much of their non-petroleum investment in Vietnamese SMEs; these firms assemble electronic goods, automobiles, and motorbikes from Japanese inputs and technology (Truong and Gates, 1994). Unlike the NIE's export-orientated production in Vietnam, much of the Japanese output in Vietnam remains in the country. Whilst starting with a very cautious approach up to the mid-1990s, Japanese firms are likely to become the major investors in Vietnam by the turn of the century. The firms from the Asian NIEs aim at using the cost advantage of Vietnam by transferring labour-intensive industries to a country where wages are low, and workers are disciplined and potentially highly productive (Truong and Gates, 1994).

The share of the 15 EU Member States (EU15) in Vietnam's total registered capital over the 1988–98 period is around 10 per cent. The European share reaches 12.0 per cent of the total, if Switzerland is included. The low share of FDI in Vietnam provided by the European Union mirrors the situation elsewhere in ASEAN and in the Asian region, with the notable exception of Singapore – see Table 9.2.

Table 9.2   EU Foreign Direct Investment Stock in Selected Asian Countries at end 1995[1]

	Total FDI Stock *(million ecu)*	*FDI Stock per capita* *(ecu)*
Thailand	2059	34.9
Malaysia	4177	208.9
Indonesia	1469	7.5
Singapore	10708	3824.3
Philippines	1794	26.4
Vietnam	2900	38.6
China	2322	1.9

Note: (1) The figures represent the book value of the FDI stock. The data for Spain, Greece, and Luxembourg are estimated. The data for Italy are based on adjusted cumulative flows.

Source: EUROSTAT. For Vietnam, authors' calculations based on figures from the Ministry of Planning and Investment, Vietnam.

This low share is at least partly due to the EU firms' failure to capitalize on the emerging markets in the ASEAN region and elsewhere
in East and Southeast Asia.

One contributory reason to this failure was the inadequate EU
recognition of the significant relaxation of restrictive practices
which has occurred throughout much of East and Southeast Asia
(UNCTAD, 1996). Furthermore, the burden of transactions costs
appears to have been highest for EU investors. Japanese firms have
set up institutional bodies in order to provide information on outward investment to Vietnam and other Asian economies. A similar
level of institutional assistance has not been provided for either EU
or US firms (UNCTAD, 1996). In general, the European Union has
relied more on trade with the region. EU exports to the ASEAN
countries stood at more than 30 billion ecu in 1994, whereas EU FDI
flows to the region did not exceed 3 billion ecu in the same year
(EUROSTAT, 1996, 1998).

However, although this low share confirms the timid involvement
of EU firms in Asia in general, it is worth noting that EU firms out-
perform their American counterparts in the case of Vietnam. This
relatively favourable performance of EU firms is explained by the
prominent position assumed by French investors on the Vietnamese
market. French firms represent more than 40 per cent of cumulative
European FDI. The Netherlands, Switzerland, and Germany have
also been amongst the most dynamic European investors in Vietnam,
albeit to a much lesser degree.

Because of the historical ties developed between Vietnam and
France, the French authorities have seen in Vietnam's *Doi Moi* policy
an opportunity to re-introduce and affirm the role of the French
language and culture in the Southeast Asian region.[8] The French
Government has demonstrated a keen interest in reviving a new form
of partnership with its old colony, notwithstanding the lethargic
economic state in which eighty years of French rule left the country.
The entry and operation of French firms has been greatly facilitated
by the many different services offered by the *Postes d'Expansion
Economique*, an administrative structure attached to the French
Ministry of Economy, Finance and Industry, which has been established in both Hanoi and Ho Chi Minh City. In addition to the
establishment of four French banks, it is estimated that there are
around 100 offices representing French firms in Vietnam. As a

result, more than one hundred investment projects with a cumulative value of US$1700 m make France the sixth biggest investor in Vietnam, before both the United States and the United Kingdom. Alcatel, TOTAL, Suez-Lyonnaise des Eaux, Sanofi, Roussel Uclaf et Rhône Poulenc, and France Câbles are among the biggest French firms with projects in Vietnam.

Following the publication of the European Commission strategy paper (Commission of the European Communities, 1994) indicating the urgent need for EU officials and businesses to help stimulate EU firms' economic presence in Asia, the Commission has strengthened partnerships with many of the Asian countries. In particular, the Commission signed a cooperation agreement with Vietnam which took effect from June 1996. The agreement deals with cooperation, economic development, trade, and investment. The protocol extending the EC–ASEAN agreement to Vietnam was signed in February 1997 (Commission of the European Communities, 1997). It is expected that Vietnam will benefit from European involvement via capital injection, technology, management, and organizational knowledge. In turn, the EU will achieve long-term involvement in this dynamic country and region.

The Major Industrial Sectors

As seen in Table 9.3, the manufacturing sector has attracted one-third of the total FDI commitments over the period 1988–98. The building and civil engineering sector makes up 29.1 per cent of the total. Dominated by hotels and tourism development, the services sector represents another 27.4 per cent. By contrast, the natural resource-based industries, and in particular agriculture, received very little foreign investment (less than 9.3 per cent in total).

Although during the period 1988–90, the oil and gas, forestry and fisheries, hotel and tourism sectors attracted most of the FDI into Vietnam, investment in the manufacturing sector has increased dramatically since the early 1990s. A notable change over previous trends is the increase in projects – mostly labour-intensive – classified in the 'light industry' category. Vietnam stands ready to take over the garment, footwear, and textile industries from Taiwan, Singapore, and Indonesia. The 1991–95 period also witnessed quality changes in FDI activities, with a concentration of projects in the electronics and

Table 9.3   Foreign Direct Investment in Vietnam at end October 1998,
             by Sector[1]
             (US$ m)

Sector	Number of Projects	Total Capital
Manufacturing		
Heavy Industry	463	5300
Light Industry	636	3600
Food Processing	155	1900
Natural Resource-based		
Oil and Gas	53	1700
Agriculture & Forestry	226	995
Maritime Products	87	330
Services		
Hotels, Tourism	190	4900
Telecommunications, Posts	128	2800
Services	104	582
Health Services	79	448
Banking & Finance	28	193
Building and Civil Engineering		
Urban Infrastructure	3	3300
Construction	248	3300
Office Property	104	2900
Others[2]	9	329

Notes: (1) Cumulative FDI over the period from 1 January 1988 to 31
             October 1998. The figures relate to committed investment,
             or to investment for which licences have been approved and
             officially registered. Only a third of the committed
             investment has been effectively disbursed.
             (2) This category includes mostly investment in EPZs and IZs.
             Source: Ministry of Planning and Investment, Vietnam.

motor vehicle industries.[9] It was during this period that Export Processing Zones (EPZs) and IZs (Industrial Zones/Parks) were set
up, facilitating increased employment within the newly-established
industries. In line with policies developed by other countries in the
region, the Vietnamese Government introduced in 1991 a set of
regulations facilitating the establishment of EPZs. By 1996, six EPZs
had been licensed, including two in the Ho Chi Minh City region.
Foreign and domestic companies can invest in EPZs in the three

major forms of investment as stipulated by the FIL, but foreign-invested projects in these areas are restricted to export-oriented businesses. Incentives available to investors in these zones are more attractive than those offered elsewhere in the country. In particular, production enterprises are subject to a 10 per cent tax rate, and are exempted from taxes during four years commencing with the first profit-making year; and all enterprises in the EPZs are exempt from duties on raw materials, spare parts etc. (Schot, 1996). The establishment of the EPZs was also an attempt at simplifying procedures for foreign investors, by providing a distinct geographical area where all services related to foreign investment were supplied. The creation of the IZs was of more recent origin (1995). Seen as the backbone of the future economy, IZs are industrial areas where both Vietnamese and foreign-invested enterprises can establish production and service projects for both export and domestic consumption. Better infrastructure and more efficient approval procedures should make these IZs more attractive than the EPZs.[10] The ultimate objective of the IZs is to encourage the development of different types of high-technology enterprises and of Research Centres.

During the second half of the 1990s, projects were being encouraged in areas such as oil, cement, telecommunications, seaports and airports, electronics, energy, steel production, chemicals, and high-tech goods.[11] This new wave of FDI projects in manufacturing, construction, telecommunications, transport, and infrastructure was credited with providing a better momentum for economic development.[12] Large investments were also committed to established production facilities within the EPZs and IZs (IMF, 1996).[13]

The Regional Distribution of FDI within Vietnam

Increasing income inequalities during the take-off stage of economic development in a given country is a phenomenon that has been observed in many developing countries, and that has been studied at length (Kuznets, 1955). According to Harrod–Domar models, divergence in welfare levels among regions will occur given initial differences in regional growth rates. For a developing nation, the divergence effect is dominant in the early stage of economic development. Convergence effects supersede divergence effects when the economies reaches higher levels of development.

As with other Asian economies (e.g. Thailand and Mainland China), the opening of the country to foreign investors has aggravated regional disparities that were rooted into the country's historical economic development. FDI was initially concentrated exclusively in the more developed provinces in the Red River Delta, and in the Northeast of Southland region in the south of the country[14] – see Figure 9.1. The Mekong Delta accounted for the remainder of the FDI projects in the south (Truong and Gates, 1994). In the North Mountains and Midland regions of the north, foreign investment projects have been rare; the exception being the growth triangle represented by Hanoi–Hai Phong–Quang Ninh.[15]

Moreover, the urban–rural divide appears clearly in the case of Vietnam, with the bulk of investment going to the industrial centres (where less than 20 per cent of the population lives) of Ho Chi Minh City and Hanoi/Haiphong, a port city which has a direct link to the industrial zones (IMF, 1996). Ho Chi Minh City's success is derived from its experience in dealing with foreigners and with Western-style business practices, along with its more developed physical and human resources. It is probable that these industrial centres and towns situated along the coastline will continue to attract the lion's share of FDI, until Vietnam's infrastructural network is upgraded (Truong and Gates, 1994).

However, escalating costs in the South and increased cross-border trade opportunities with China, combined with a careful regional policy, have all started to redress the regional imbalance in the country. The promulgation of preferential tax incentives in October 1992 was a first step towards the redistribution of foreign investment in favour of the poorer northern and central provinces. In 1991, a growth pole strategy (with the creation of EPZs) was being developed in order to allow growth effects to diffuse out of the urban areas and towards the poorer rural areas of the country (Nestor, 1997). As a result, several poorer localities have been able to attract foreign investment for the first time: these include Lai Chau, Ha Giang, Yen Bai in the Northern Mountains and Midlands; Quang Ngai on the South Central Coast; and Gia Lai in the Central Highlands.[16] The concentration of mineral deposits and poor infrastructure in the North may, however, accentuate the regional specialization of FDI projects, preventing the Northern regions from moving into manufacturing and consumer goods industries.

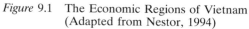

Figure 9.1   The Economic Regions of Vietnam
(Adapted from Nestor, 1994)

Provinces:

01 Hanoi
02 Ho Chi Minh City
03 Hai Phong
10 Ha Giang
11 Cao Bang
12 Lai Chau
13 Lao Cai
14 Tuyen Quang
15 Lang Son
16 Bac Thai
17 Yen Bai
18 Son La
19 Vinh Phu
20 Quang Ninh
21 Ha Bac
22 Ha Tay
23 Hai Hung
24 Hoa Binh
25 Nam Ha
26 Thai Binh
27 Thanh Hoa
28 Ninh Binh
29 Nghe An
30 Ha Tinh
31 Quang Binh
32 Quang Tri
33 Thua Thien-Hue
34 Quang Nam-Da Nang
35 Quang Ngai
36 Kon Tum
37 Binh Dinh
38 Gia Lai
39 Phu Yen
40 Dac Lac
41 Khanh Hoa
42 Lam Dong
43 Song Be
44 Ninh Thuan
45 Tay Ninh
46 Binh Thuan
47 Dong Nai
48 Long An
49 Dong Thap
50 An Giang
51 Ba Ria-Vung Tau
52 Tien Giang
53 Kien Giang
54 Can Tho
55 Ben Tho
56 Tra Vinh
57 Vinh Long
58 Soc Trang
59 Minh Hai

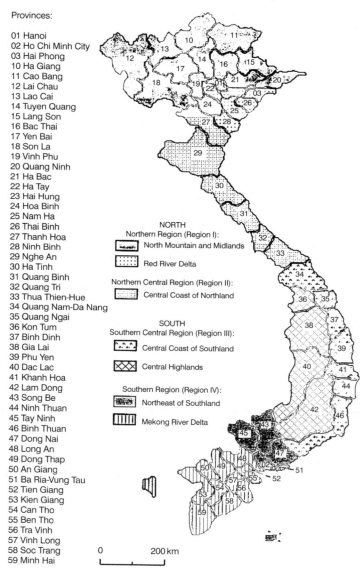

NORTH
Northern Region (Region I):
North Mountain and Midlands

Red River Delta

Northern Central Region (Region II):
Central Coast of Northland

SOUTH
Southern Central Region (Region III):
Central Coast of Southland

Central Highlands

Southern Region (Region IV):
Northeast of Southland

Mekong River Delta

0   200 km

Although most foreign investors find Hanoi people more educated than their southern counterparts, the Southern regions have been marked by a strong Chinese work ethic, rooted in the trade monopolies granted to them by the French in the 19th Century and onwards. But differences in history and geography between the north and the south are shaping different trade and investment relationships. The south of the country is creating strong relationships with Taiwan, Singapore, South Korea, and Japan, whereas the north is developing tighter commercial ties with the southern provinces of Mainland China. For example, Taiwan, which is Vietnam's largest investor, has invested almost exclusively in the South (EIU, 1996).

THE CHOICE OF LOCATION WITHIN VIETNAM

From the standpoint of the transnational corporation, the choice of location within a country is one of the important decisions that determine the performance of its foreign affiliate (Dicken, 1998). Equally, the TNC's location decision will be significant for the host country, in terms of its national and regional economic and social development. When the decision is made to locate overseas rather than export or license, the TNC faces two interrelated decisions;[17] it needs to decide upon (i) a suitable host country within an overall region, and (ii) a specific province within the host country to invest in. A substantial number of empirical studies have identified the variables, both economic and institutional, which influence the TNC's choice of a given host country: labour costs (Kravis and Lipsey, 1982; Yamawaki, 1993); the availability of skilled labour (Hasegawa, 1996; Ó hÚallacháin and Reid, 1992); market proximity (Thomsen, 1993); financial assistance (Taylor, 1993), and the existence of a business infrastructure[18] (Dunning, 1993; Yamawaki, 1993). The level of bureaucracy, socio-political stability, and other risk factors have been introduced in some econometric studies.[19]

Various empirical studies have put forward a multitude of variables explaining the attractiveness of specific regions within a given country (Little, 1978; Hill and Munday, 1992; Glickman and Woodward, 1988; Arpan, 1981). Underlying these variables is the concern of the foreign investor for cost minimization as well as for access to key markets. The variables which have been established to have a significant effect upon the spatial distribution of FDI within a country are:

- The availability of natural resources, together with labour and other production costs. Most studies have found a negative relationship between investment and labour costs (see, for example, Thiran and Yamawaki, 1995), although Swedenborg (1979) found that Swedish FDI goes to countries with high labour costs, presumably to exploit the technological capability of the host county.
- Market proximity and population density (Sekiguchi, 1979; Netherlands Economic Institute, 1992). Proximity to a large market allows the firm to exploit marketing and distribution economies of scale. Proximity to airports and seaports, as well as a good quality transport network, facilitate market access: such considerations are crucial in the case of export-oriented investments. Membership of an economically integrated region, such as a free trade area or a customs union, represents an additional incentive for foreign firms wishing to export to neighbour member countries. This consideration is particularly important for Japanese investors who have successfully built up vertically-integrated production processes within the ASEAN area.
- Agglomeration effects related to the density of manufacturing activity in a given province or region. These effects include clusters of potential customers, and of reliable and good-quality suppliers. They also refer to the effects of other firms operating in related areas of activity, such as those found in a typical Italian industrial district or in a typical Asian EPZ. Indeed, Wheeler and Mody (1992) suggest that agglomeration economies have a dominant influence on investors' calculations.
- The business infrastructure, in particular whether a type of industrial activity benefits from fiscal incentives, and whether an area has assisted area or zoning status and receives financial assistance (Taylor, 1993; Hill and Munday, 1992).
- The technological capability of regions, and the skills of the workforce (Cantwell, 1989; Andréosso-O'Callaghan, 1997). The benefits to the TNC accrue from its ability to tap into the technological specialization of different locations (Young *et al.*, 1994).
- Unemployment is another variable that has been considered in the explanation of FDI flows. High unemployment rates signal

the availability of labour (Coughlin *et al.*, 1991; Thiran and Yamawaki, 1995).

In order to test the relative importance of the variables in the locational decision of firms, the following multiple regression model was estimated:

$$\text{FDI}_{it} = \alpha + \beta_1\text{GDP}_{it} + \beta_2\text{TRANS}_{it} + \beta_3\text{INC}_{it} + \beta_4\text{EDUC}_{it} + \varepsilon_{it}$$

FDI_{it} is the legal capital invested (in US\$'000) in region *i* in period *t*, divided by the total population of region *i* in period *t*. GDP_{it} is Gross Domestic Product per capita in region *i* in period *t*, and gives an indication of income across regions. We expect those foreign firms in search of new markets in Vietnam to be attracted to the regions with higher per capita incomes, in which case the sign of the estimated coefficient should be positive. On the other hand, since GDP per capita figures also reflect labour cost differentials, a predominance of cost-minimizing firms would give rise to a negative sign. The sign of the coefficient is thus indeterminate *a priori*.

TRANS_{it} gives an indication of communication links between the foreign-invested enterprise and the market (either domestic or foreign). It is measured by the importance of freight in region *i*, and is calculated as million tons of freight per square kilometre of land. We expect that inward investment will be positively related to TRANS, since large volumes of freight reflect intense economic activity in the region. This, in turn, suggests the existence of a relatively developed transportation infrastructure, and also the existence of a substantial market.

INC_{it} refers to the availability of financial incentives such as those found in export processing zones. It is a dummy variable which takes a value of unity if there is an EPZ or an IZ in a given region, and a value of zero otherwise. EDUC_{it} refers to the educational level in region *i* and is measured as the number of pupils/students in technical training schools, technical secondary schools, and in universities and colleges, divided by the total population in each region.

We use a pooled time-series and cross-section regression model. Data have been collected for seven regions, for each of the years 1993–95. We assume that FDI in year *t* is made on the basis of information gathered in year *t* − 1, hence one-year lagged values of

Table 9.4 The Determinants of Foreign Direct Investment within Vietnam: Regression Results

Explanatory Variable	Coefficient	t-statistic
GDP per capita	1.10	3.26
Transportation	−0.0012	0.53
Education level	89.17	1.9
Incentives	−16.19	0.74
Constant	−39.7	1.8

Durbin Watson = 2.7226.
$R^2 = 0.69$.
Adjusted $R^2 = 0.61$.
Sample size $n = 21$.

Notes: (1) Annual data for seven provinces over the period 1993–95.
(2) The Durbin Watson test for autocorrelation was inconclusive.
(3) chi-Square = 6.7, with 4 degrees of freedom.
(4) A sequential Chow Test indicated there was no structural change between regions.

the independent variables are used in the model. The small size of the sample is unavoidable given the lack of additional data, but does mean that there are not many degrees of freedom in the regression model and hence that it may be difficult to achieve statistically significant results.

Nevertheless the model does shed some light on the major determinants of FDI in Vietnam. As can be seen in Table 9.4, the value of the coefficient of determination (R^2) implies that 69 per cent of the variation in FDI at the intra-national level is explained by the four independent variables in the regression model.

Our core result shows that GDP is a significant determinant of cross-regional FDI flows in Vietnam at the 5 per cent level of significance. The positive sign of the regression coefficient means that FDI is attracted to the richer provinces, notwithstanding the fact that labour costs may also be higher there. This is consistent with our observation that FDI is essentially concentrated in the more affluent urban centres, as opposed to the less wealthy rural areas. This implies that although foreign investors may well be attracted to Vietnam because of its superior (labour) cost advantages when compared with other Southeast Asian nations,[20] inter-provincial labour costs differentials do not matter much.

Our results allow us to substantiate the hypothesis that the deliberate policy of redistributing FDI on a spatial scale, through the creation of EPZs and IZs and with the provision of financial incentives (INC), has not yet born fruit. The coefficients of both TRANS and INC are insignificant and, in particular, the effect of good communication links (TRANS) appears to be totally negligible. Finally, the EDUC variable is found to be significant at the 10 per cent level suggesting that a well-educated workforce is an additional attraction to investors. At the national level, it is interesting to note that Vietnam's illiteracy rate (6.3 per cent in 1995) is similar to that in Thailand (6.2 per cent), but well below China (18.5), Malaysia (16.5 per cent), Indonesia (16.2), and also Singapore (8.9 per cent in 1995).

CONCLUDING REMARKS

When compared with other Asian economies, Vietnam is a late opener to FDI, and the country is consequently still learning and adapting its reform process. After a brief initial emphasis on the development of offshore oil and gas deposits, FDI flows in Vietnam have increasingly concentrated on the manufacturing sector (garments and textiles, footwear, high-technology goods, motor vehicles, construction), as well as telecommunications. In line with the situation recorded in other Asian economies, most of the FDI inflows into Vietnam originate from the Asian newly industrialized economies (Taiwan, Hong Kong, South Korea), and from other ASEAN countries, particularly Singapore. Since the lifting of the US embargo in 1994, Japanese investment has risen considerably, and Japanese investors are the major source of foreign-invested projects in Vietnam at the turn of the century. French investors represent more than 40 per cent of the EU FDI share in Vietnam, a pattern of relations which has been shaped greatly by the ties existing between the two countries in the past. Although the effects of the Asian crisis have not been fully felt in Vietnam, the renewed interest in the

Vietnamese economy on the part of French firms has provided a cushioning mechanism.

The results of our multiple regression model indicate that both GDP per capita and educational level (EDUC) have had an important influence on the location of FDI within Vietnam. The fact that FDI flows are attracted by regions where GDP and educational levels are relatively high, suggests that the uneven regional distribution of FDI will continue. The majority of FDI projects will go to the established industrial centres, accentuating the urban–rural divide. Geographically, FDI is highly concentrated in the Northeast of Southland, parts of the Central Coast of Northland and Southland, and the Red River Delta region bordering the North Mountain and Midlands. Vietnamese planners have begun to recognize this inequality and have started to prioritize certain regions to reduce this divide. However, the establishment of EPZs in the early 1990s and of IZs later on, has not yet produced the expected diffusion of FDI activity. Our model shows that the impact of communication links and of incentives – embodied in the EPZs and IZs – on the intra-national distribution of FDI is negligible. It follows that a regionally-balanced growth policy in the future requires further domestic and international investment in both physical and human capital formation.

NOTES

1. The authors would like to extend their great appreciation and thanks to many individuals for their guidance in the preparation of this paper, in particular: Bill Turley, Adam Fforde, Curt Nestor, Ray Feddema, Ray Mallon, Ari Kokko, VietVu, Prof. TT Nguyen (VECON), Thanh-Dam Troung, Chris Moore, Mariama Williams, and Synnøva O'Gorman.
2. See for example a recent contribution by Rodriguez-Clare (1996).
3. See the *Vietnam Investment Review 1988*.
4. See the very incisive chapter by Chee Peng Lim (1995) on the possible complementarities existing between Vietnam and other ASEAN members.
5. For an insight into the main amendments of the FIL, see Mason (1998).
6. Although the diffusion of the crisis has not fully hit Vietnam, the optimistic growth rate target of 9 per cent for 1998 had to be adjusted downwards to around 6 per cent (Kokko, 1998).

7. The case of China may be perhaps the more appropriate comparison, since the economic reforms only started to produce an impact on the Chinese manufacturing sector in the mid to late 1980s.

8. There is an intense campaign to revive the French language, despite the fact that it is today only spoken by approximately 1 per cent of the Vietnamese population (Ambassade de France au Vietnamn, 1998).

9. See the reports in the *Saigon Times Daily*.

10. It is worth noting that the Industrial Zones were created mainly as a response to the lack of success of the EPZs. The Vietnamese EPZs are indeed notorious for their limited success so far, due to a number of problems such as high prices for land rents and utility services, and underdeveloped infrastructure (Schot, 1996).

11. See the reports in the *Vietnam Investment Review*.

12. See the reports in the *Saigon Times Daily*.

13. The average capital for a project increased from US$3.5 m in 1988–89 to US$16.4 m in 1995. See the *Economic Development Review* (November 1996).

14. The growth triangle of the South is referred to as the 'Southern Economic Focal Zone' (Forbes, 1996). This includes Ho Chi Minh City (in Dong Nai province), Vung Tau (North-East South) and Bien Hoa.

15. During the period between January 1988 and 30 August 1996, 782 projects (i.e. 51 per cent of the total) representing a total capital of US$11.25 bn (i.e. 48 per cent of total investment in Vietnam) were realized in the Eastern-South region (including Ho Chi Minh City, Dong Nai, Song Be, and Vung Tau). The Northern economic zone (including Ha Noi and Hai Phong) had attracted 292 projects with a total capital of US$5.54 bn. The provinces in Central Vietnam (from Nghe An to Lam Dong) attracted 108 projects (7 per cent) with total capital of US$1165 m (5 per cent). These data suggest that those provinces with good infrastructure and reliable supplies of labour have been most likely to attract foreign investment. *Economic Development Review* (November 1996).

16. See the *Saigon Times Daily* 1998.

17. Thiran and Yamawaki (1995) suggest that Japanese investors tend to evaluate simultaneously country-specific and region-specific factors in their location decisions.

18. For example, research and development capabilities (overall technological infrastructure), educational facilities, and the legal/financial framework.

19. For example, Shah and Slemrod (1990) introduced a general index of Mexico's creditworthiness in their model.

20. The monthly labour cost rate in a manufacturing firm is US$35 in Vietnam, compared with for example US$96 in Bangkok and US$77.2 in the remote provinces of Thailand (Ambassade de France au Vietnam, 1998; and Thai Board of Investment, Bangkok).

BIBLIOGRAPHY

Ambassade de France au Vietnam (1998) *Le Vietnam et la co-opération Décentralisée Française* (Hanoi: Ambassade de France au Vietnam).

Andréosso-O'Callaghan, B. (1996) 'The Spatial Impact of Japanese Direct Investment in France'. In J. Darby (ed.), *Japan and the European Periphery*, pp. 111–31 (London: Macmillan).

Andréosso-O'Callaghan, B. (1997) 'Japanese Manufacturing Investment in France in a EU Comparative Framework: Theory and Practice'. *Proceedings of the International Conference on 'Japan and the Mediterranean Countries'*, Università Degli Studi di Napoli.

Apoteker, T. (1996) *Vietnam – Supplementary Analysis on Foreign Direct Investment*. Consultant Report, Analyses économiques et financières, Université de Rennes, France.

Arpan, J. (1981) 'The Impact of State Incentives on Foreign Investors' Site Selections', *Economic Review*, 66, 8, 36–42.

Booth, A. (1997) 'Vietnam and ASEAN: How far apart?' *Proceedings of the Conference on 'Vietnam Reform and Transformation'*, Centre for Pacific Asia Studies, Stockholm University.

Cantwell, J. (1989) *Technical Innovations in Multinational Corporations*. (London: Basil Blackwell).

Chee Peng Lim (1995) 'ASEAN–Indochina Relations: Prospects and Scope for Enhanced Economic Co-operation'. In T. Kawagoe and S. Sekiguchi (eds), *East Asian Economies: Transformation and Challenges*, pp. 305–31 (Singapore: Institute of Southeast Asian Studies).

Collis, C. and Noon, D. (1993) 'Foreign Direct Investment in the UK Regions: Recent Trends and Policy Issues', *Regional Studies*, 28, 8, 843–8.

Commission of the European Communities (1994) *Towards a New Asia Strategy*. COM(94) 314 final (Luxembourg: Office for Official Publications of the European Communities).

Commission of the European Communities (1997) *General Report on the Activities of the European Union* (Luxembourg: Office for Official Publications of the European Communities).

Coughlin, C.C., Terza, J.V. and Arromdee, V. (1991) 'State Characteristics and the Location of Foreign Direct Investment within the United States', *Review of Economics and Statistics*, 73, 4, 675–83.

Dicken, P. (1998) *Global Shift*, Third Edition (London: Paul Chapman).

Dunning, J.H. (1993) *Multinational Enterprises and the Global Economy* (New York: Addison-Wesley).

Economic Development Review (1996) November (Ho Chi Minh City: University of Economics).

Economist Intelligence Unit (1996) *Country Report: Vietnam* (London: EIU).

Embassy of Vietnam (1998) 'Vietnam–EU Relations Further Strengthened'. Press release (Washington DC: Embassy of Vietnam).

EUROSTAT (1996) *Statistics in Focus: External Trade* (Luxembourg: Office for Official Publications of the European Communities).

EUROSTAT (1998) *Key Figures: Relations between the European Union and Asian ASEM countries* (Luxembourg: Office for Official Publications of the European Communities).

Ezaki, M. and Son Le Anh (1997) 'Prospect of the Vietnamese Economy in the Medium and Long Run: A Dynamic CGE Analysis'. APEC Discussion Paper Series no. 10. Nagoya University & Institute of Developing Economies, Graduate School of International Development, APEC Study Center.

Forbes, D. (1996) 'Urbanisation, Migration, and Vietnam's Spatial Structure', *SOJOURN*, 11, 1, 24–51.

Glickman, N. and Woodward, D. (1988) 'The Location of Foreign Direct Investment in the US: Patterns and Determinants', *International Regional Science Review*, 11, 137–54.

Hasegawa, S. (1996) 'Determinants of Japanese Firms' Entry Modes in Europe', *Colloquium on 'Globalisation et Régionalisation'*, Université Panthéon-Sorbonne, Paris.

Hill, S. and Munday, M. (1992) 'The UK Regional Distribution of Foreign Direct Investment: Analysis and Determinants', *Regional Studies*, 26, 6, 535–44.

Hirschman, O. (1958) *The Strategy of Economic Development* (New Haven CT: Yale University Press).

Imada, P., Montes, M. and Naya, S. (1991) *A Free Trade Area: Implications for ASEAN* (Singapore: Institute of Southeast Asian Studies).

International Monetary Fund (1996) 'Vietnam: Transition to a Market Economy', IMF Occasional Paper 135 (Washington DC: IMF).

Kokko, A. (1998) 'Vietnam: Ready for Doi Moi II?' Stockholm School of Economics, Economic Research Institute Working Paper no. 286.

Kolko, G. (1997) *Vietnam: Anatomy of a Peace* (London: Routledge).

Kravis, I.B. and Lipsey, R.E. (1982) 'The Location of Overseas Production and Production for Export by US Multinational Firms', *Journal of International Economics*, 12, 3/4, 201–23.

Kuznets, S. (1955) 'Economic Growth and Income Inequality', *American Economic Review*, 45, 1, 1–28.

Little, J.S. (1978) 'Locational Decisions of Foreign Direct Investors in the US', *New England Economic Review*, July/August, 43–63.

Ljunggern, B. (1997) 'Vietnam's Second Decade under Doi Moi: Emerging Contradictions in the Reform Process', *Proceedings of the Conference on 'Vietnam Reform and Transformation'*, Centre for Pacific Asia Studies, Stockholm University.

Mason, M. (1998) 'Foreign Direct Investment in Vietnam: Government Policies and Corporate Strategies', *EXIM Review*, 17, 2, 1–70.

McConnell, E. (1990) 'Foreign Direct Investment in the US', *Annals of the Institute of American Geography*, 70, 259–70.

McDermott, P.J. (1977) 'Overseas Investment and the Industrial Geography of the UK', *Area*, 9, 200–7.

Molteni, C. (1997) 'Japanese Direct Investments in Europe: The Italian Case', *Proceedings of the International Conference on 'Japan and the Mediterranean Countries'*, Università Degli Studi di Napoli.

Nestor, C. (1994) 'Foreign Joint Ventures in the Socialist Republic of Vietnam 1988–1993'. Unpublished dissertation. Department of Human and Economic Geography, School of Economics and Commercial Law, Gothenburg University.

Nestor, C. (1997) 'Foreign Investment and the Spatial Pattern of Growth in Vietnam'. In C. Dixon and D. Drakakis-Smith (eds), *Uneven Development in South-East Asia* (New York: Ashgate).

Netherlands Economic Institute (1992) *New Location Factors for Mobile Investment in Europe* (Rotterdam/London: NEI/Ernst and Young).

Ó hUallacháin, B. and Reid, N. (1992) 'Source Country Differences in the Spatial Distribution of Foreign Direct Investment in the United States', *Journal of the Professional Geographer*, 44, 3, 272–85.

Reidel, J. (1997) 'Transition to a Market Economy in Vietnam'. In W.T. Woo, S. Parker and J. Sachs (eds), *Economies in Transition: Comparing Asia and Eastern Europe* (London: MIT Press).

Rodriguez-Clare, A. (1996) 'Multinationals, Linkages, and Economic Development', *American Economic Review*, 86, 4, 852–72.

Saigon Times Daily, various issues.

Schot, A.C.M. (1996) *Legal Aspects of Foreign Investment in the Socialist Republic of Vietnam* (London: Kluwer Law International).

Sekiguchi, S. (1979) *Japan Direct Foreign Investment* (London: Macmillan).

Shah, A. and Slemrod, J. (1990) 'Tax Sensitivity of Foreign Direct Investments: an Empirical Assessment'. World Bank Policy Research and External Affairs, Working Paper no. 434.

Swedenborg, B. (1979) 'The Multinational Operations of Swedish Firms: an Analysis of Determinants and Effects' (Stockholm: The Industrial Institute for Economic and Social Research).

Taylor, J. (1993) 'An Analysis of the Factors Determining the Geographical Distribution of Japanese Manufacturing Investment in the UK, 1984–91', *Urban Studies*, 30, 7, 1209–24.

Thiran, J.M. and Yamawaki, H. (1995) 'Regional and Country Determinants of Locational Decisions: Japanese Multinationals in European Manufacturing', Louvain Economics Papers.

Thomsen, S. (1993) 'Japanese Direct Investment in the European Community: the Product Cycle Revisited', *The World Economy*, 16, 3, 301–15.

Tran Van Hoa (1997) 'Vietnam's Recent Economic Performance and its Impact on Trade and Investment Prospects'. In Tran Van Hoa (ed.), *Economic Development and Prospects in the ASEAN* (London: Macmillan).

Truong Dinh Hung and Gates, C. (1994) 'FDI in Vietnam: Its Role and Prospects', *Vietnam Social Sciences*, 2–94.

United Nations Conference on Trade and Development (1996) *Investing in Asia's Dynamism: European Union Direct Investment in Asia* (New York: United Nations).

US Embassy (1996) *Country Commercial Guide: Vietnam* (Hanoi: US Embassy).

Vietnam Investment Review (The Vietnam Business Journal), various issues.

Wheeler, D. and Mody, A. (1992) 'International Investment Location Decisions: the Case of US Firms', *Journal of International Economics*, 33, 57–76.

Yamawaki, H. (1993) 'Location Decisions of Japanese Multinational Firms in European Manufacturing Industries'. In K.S. Hughes (ed.), *European Competitiveness*, pp. 11–28 (Cambridge: Cambridge University Press).

Yannopoulos, G. and Dunning, J.H. (1976) 'Multinational Enterprises and Regional Development: an Exploratory Paper', *Regional Studies*, 10, 389–99.

Young, S., Hood, N. and Peters, E. (1994) 'Multinational Enterprises and Regional Economic Development', *Regional Studies*, 28, 7, 657–77.

Section IV
The Future

10 Enlargement to Include Formerly Centrally Planned Economies: ASEAN and the European Union Compared

Richard Pomfret

INTRODUCTION

During the 1990s, both the Association of Southeast Asian Nations (ASEAN) and the European Union (EU) have been preparing for enlargement to include formerly centrally planned economies. This chapter analyses the economic consequences from a comparative perspective, and concludes with an assessment of the implications of enlargement for ASEAN–EU economic relations.

In both cases the new members are significantly poorer than the pre-existing members, with a more rural economy, run-down infrastructure and often with less well-established institutional and political regimes. Integration questions are generally more pressing for the European Union with its more tightly-knit economic and political structure, and the budgetary implications of enlargement are substantial. The income gap is not so large in Southeast Asia, and ASEAN is a much looser organization than the European Union. Integration considerations are, however, not absent from ASEAN whose tradition of consensus could be threatened by enlargement to include members with differing political histories, institutions and aspirations – a point underlined by the last-minute failure of Cambodia to be admitted after the July 1997 coup.[1]

The main parts of this chapter will focus on two differences between the EU and ASEAN cases. First, the East Asian economies in transition from central planning have been more dynamic than the East European transition economies, and did not suffer the major income losses experienced by Eastern European countries during the first half of the 1990s. The second section discusses differences between Asian and European models of transition, and analyses the implications for ASEAN. The second difference to be analysed is the contrast between the Eastern European countries joining their main market and the Indochina countries joining a group of competing producers. Although the Eastern European economies are industrialized, they are potentially complementary to the existing EU countries, with the opportunity for them to specialize in agricultural products, textiles and some standardized industrial goods if they become fully integrated into the European Union. In contrast, the Indochina countries and Myanmar are actual or potential competitors of the original ASEAN members, supplying food, raw materials, and labour-intensive manufactures to markets outside Southeast Asia.

The speed and nature of enlargement of the two regional organizations will be determined by how the above issues are settled. The final section of the chapter draws some conclusions, and relates the outcomes to prospects for the future development of ASEAN–EU economic relations. A key issue is the uncertainty about where exactly comparative advantage lies in economies which were characterized by gross resource misallocation until recently. The shift to world prices implied by full membership of the European Union or ASEAN could lead to trade surges as demand for previously repressed imports booms or as previously unidentified areas of comparative advantage are exploited.

ENLARGEMENT OF ASEAN AND THE EUROPEAN UNION

The evolution of the European Union and of ASEAN up to the end of the 1980s is well-known. The European Union had developed a much deeper economic integration with a customs union and a high degree of factor mobility, plus common policies towards areas such as agriculture. ASEAN meanwhile was a much looser organization, whose initiatives towards establishing internal free trade or common policies had hardly progressed, but which had been a force for the regional stability underpinning its members' rapid growth. The

European Union had expanded from six to nine members in the 1970s, and then to 12 in the 1980s (and to 15 in 1995), reaching close to its natural limit of European market-oriented economies.[2] ASEAN had expanded from five to six members, but it too represented the natural limit of market-oriented economies in Southeast Asia in the 1980s.[3]

The unexpectedly sudden collapse of Communism in Europe, and the transition from central planning to market-based economies ushered in a new set of applicants for EU membership in the 1990s. The Central and East European economies saw the European Union as an attractive partner for political as well as economic reasons, reaffirming their historic role as Europeans and providing a seal of approval for the adoption of democratic political systems. A similar change happened in Southeast Asia, as Vietnam, Cambodia, Laos, and Myanmar shifted from earlier antagonism towards ASEAN to a desire for membership. As in Europe, political considerations played a role as Vietnam no longer saw the ASEAN countries as members of a foreign-led anti-Communist alliance but rather as natural allies against a resurgent China, and as long-isolated regimes in Cambodia and Myanmar sought international respectability. Although the political dimensions are important, this chapter will just focus on the economic consequences of enlarging the European Union and ASEAN to include economies in transition from central planning.

What is difficult to ignore in any comparison is the future limits of the European Union, or just how many of the transition economies will join, and when. In the remainder of the chapter, it will be assumed that the likely pool of potential members contains Poland, Hungary, and the Czech and Slovak Republics, as well as Bulgaria, Romania, and Slovenia, and the three Baltic republics (Estonia, Latvia, and Lithuania) so that quantitative comparisons are between the current EU15 and a putative EU25; this assumption makes no prediction about what will happen but the argument is little affected by whether the smaller economies, in particular the last four listed, are in or out.[4] The before and after comparison is more straightforward for ASEAN, since of the SEA-10 only Cambodia is not yet an ASEAN member and its accession would not have a major economic impact.

Table 10.1, based on Langhammer (1997b: 4), illustrates some similarities and differences. EU enlargement from 15 to 25 members

Table 10.1   Selected Statistics to Illustrate the Effects of Possible Enlargements of the European Union and the Association of Southeast Asian Nations

	% Increase as a Result of Enlargement	
	EU15→ EU25	*ASEAN6→SEA-10*
Population	28	37
Land Area	34	47
Gross National Product	4	6
Agricultural Labour	55	45

Notes: EU15 refers to the 15 Member States of the European Union from 1995; EU25 refers to a potential Union incorporating 10 additional countries from Eastern Europe – see text for details. ASEAN6 refers to the six members of ASEAN before the accession of Vietnam, Laos, and Myanmar; SEA-10 refers to these nine countries plus Cambodia.
Source: Langhammer (1997b: 4).

increases population and land area by 28 per cent and 34 per cent respectively, while expanding ASEAN from six to ten members increases population by 37 per cent and land area by 47 per cent: i.e. the impact is larger by about one-third. In current dollar GNP, the comparison is between a 4 per cent increase and a 6 per cent increase; the new members are much poorer than the existing members, although the gap is smaller for ASEAN than for the European Union. Another interesting comparison is the percentage increase in the agricultural labour force, which would be 55 per cent for the European Union and 50 per cent for ASEAN; in both cases the newcomers are more agrarian than the existing members, but the percentage increase is more pronounced for the European Union. This is especially important given that agricultural and regional policies account for four-fifths of the EU budget, so that the eastern expansion is likely to require substantial EU policy reforms.[5]

In sum, the enlargement of the European Union is more problematic than ASEAN enlargement. The EU's deeper integration, especially the cost of bringing poor rural countries inside the common agricultural policy, is the major reason for slower progress. Also, the larger economic distance from Brussels to Eastern Europe than from ASEAN to Indochina (Fischer *et al.*, 1997; Langhammer, 1997a) may add to EU caution.

TRANSITION IN ASIA AND IN EUROPE

The contrasts between the Asian and the European transition economies are often perceived to be strong, although there is less agreement as to which differences are significant. All of the centrally planned economies were strongly influenced by the Soviet model, even though significant differences had emerged in China and Yugoslavia by the time Mao and Tito died. Major reforms were introduced earlier in China than in any of the Central and Eastern European countries (other than Yugoslavia). But the most striking difference between the Asian and European transition economies is in post-reform performance. Whereas all the European transition economies suffered a substantial drop in output and almost all of them experienced severe hyperinflation, both China and Vietnam enjoyed accelerated output growth after introducing reforms and China avoided high inflation.

How to explain these differences? The initial response of China specialists was to claim a superior transition strategy based on grad-ual change rather than the Big Bang adopted in Eastern European countries such as Poland (Chen *et al.*, 1992). Although this claim had some superficial plausibility, it did not stand up to wider comparative analysis. China's 1978/79 reforms were in fact dramatic: fundament-ally reforming agriculture where over four-fifths of the population worked, and permitting foreign investment for the first time since 1949.[6] On the other hand, some of the Eastern European countries were cautious about the speed of reforms, and yet all of them experienced the same initial output loss (Blanchard, 1997). Vietnam is difficult to classify, in that it followed a similar agriculture-plus-open-door strategy to China's when it began reforms in 1986, but also adopted a Polish-type strict macroeconomic policy when the reform process was revitalized in 1989.

The response of supporters of rapid transition was to explain the differing Asian performance by different initial conditions (Sachs and Woo, 1994). The main advantage of the Asian reformers was their large agricultural sector and readily identifiable comparative advantage in labour-intensive activities. As a general explanation, the 'initial conditions' view is unconvincing because some initial conditions (e.g. bigger stock of human capital) favoured the European transition economies. On the other hand, central planning was especially flawed in agriculture, with its large number of producers operating in varying natural conditions (soil, climate, etc.), and *a fortiori* in rice farming where there were not even major economies of scale to be reaped from collectivization. In industry, it was easier to release rural labour into new manufacturing activities than to change the distorted output mix of an inefficient industrial sector. Both China and Vietnam, and the smaller Indochina countries, benefited from having relatively few State enterprises to deal with; privatization and restructuring were major issues in Eastern Europe and the former USSR because State-owned industrial enterprises accounted for the majority of economic activity, while in China they employed less than one-sixth of the workforce.

A third view of the difference between the post-reform performances of Asian and European centrally planned economies is that motives varied. In Eastern Europe transition to a market-oriented economy was undertaken by new governments, often with a commitment to the market mechanism and private ownership as the underpinnings of a new political system. In brief, transition was a goal in itself or, at a minimum, an important symbol of a new era. In East Asian centrally planned economies, the Communist Party remained in power and was reluctant to even refer to 'transition' to a more market-oriented economy, instead preferring convoluted images of renewal or of new economic policies or of a social market economy. In brief, the emphasis was on continuity. This was also true of some former Soviet Republics but the difference between, say, Turkmenistan or Belarus which tried to preserve the economic status quo, and China or Vietnam, is that the leadership in China and Vietnam recognized the necessity of promoting economic growth in order to bolster their own survival chances. A growth strategy may involve a greater role for market mechanisms, but the leadership typically saw this as a price to be paid rather than a benefit in itself.

The recipe for accelerating growth in China or Vietnam was well-known from the experience of the East Asian newly industrialized economies next door. The key component was exporting labour-intensive manufactures, which involved opening up the economy to international trade and also to the transfer of skills needed to produce and sell such products in the high income countries. Improved agricultural productivity would allow resources (especially labour) to be transferred from agriculture into labour-intensive manufacturing without pressure on food supplies.[7] The entire history of Chinese economic policy since 1978 can be interpreted as working to support this growth strategy (Pomfret, 1997b). To be sure, this led to a more market-oriented economy, but measures necessary to a market economy whose worth in promoting growth was unclear were not undertaken; in the financial sector, for example, the foreign exchange market was gradually opened up but domestic asset markets remained tightly regulated and inefficient in allocating capital. Vietnam has followed China in broad outline, with similar lethargy in reforming State-owned enterprises, but greater willingness to liberalize prices.

What are the implications of this analysis for Southeast Asia? The five original ASEAN members and the four seeking membership in the 1990s were really not so different 30 years ago – see Table 10.2. Apart from Singapore and Brunei, all of the SEA-10 were low to

Table 10.2   GNP per capita for the SEA-10 Countries, 1970 and 1995

	1970	1995
Brunei	—	14240[b]
Cambodia	130	270
Indonesia	80	980
Laos	120	350
Malaysia	380	3890
Myanmar	80	235[c]
Philippines	210	1050
Singapore	920	26730
Thailand	200	2740
Vietnam	140[a]	240

Notes: (a) GDP-weighted average of the two Vietnams.
(b) 1994 data, from the *World Bank Atlas 1996*.
(c) 1994 data, from Langhammer (1997a: 163).
Sources: *World Bank Atlas 1972*; *World Development Indicators*, 1997.

middle income countries which had been drawn into the global division of labour as primary product exporters and which had not developed substantial manufacturing sectors. In the 1970s and 1980s, with centrally planned economies insulated from world markets, Vietnam, Laos, Cambodia, and Myanmar all suffered increasingly from resource misallocation and economic stagnation. The rapid growth of Malaysia, Thailand, and Indonesia, especially in the decade up to 1996, opened up a substantial income gap. The gap is, however, less wide than the one opened up between Western and Eastern Europe during 40 years of central planning in the east, and it is more readily bridgeable because the Southeast Asian transition economies can follow the growth path of the ASEAN countries.

COMPETITION AND COMPLEMENTARITY

The EU15 are more industrial than the 10 Eastern European countries wishing to join the European Union. Should not this complementarity increase the potential gains from freer trade within an EU25, as the east trades food, raw materials and basic industrial goods in return for more skill-intensive manufactures and services from Western Europe? On the other hand, if the new ASEAN members produce identical goods to the old members, will there be no gains from intra-ASEAN trade and increased friction as they compete in overseas export markets?

The complementarity of the EU15 and the European transition countries could lead to considerable trade creation as grains, coal, steel, and other Eastern European exports gain market share in the enlarged European Union at the expense of the EU15's most protected economic activities. If that were the end of the story there would be considerable gains from trade, and little trade diversion as there are few third country supplies to EU 'sensitive product' markets that could be diverted. The problem arises from the endogeneity of policy. The highly protected activities are in this situation because they are politically influential and the threat of competition from new EU members will lead them to use their political influence to prevent the trade creation outcome, which benefits consumers and exporters in the EU15 at the expense of specific factors in the protected activities.

What is the likely outcome? In agriculture, the European Union is constrained by international commitments entered into under the Uruguay Round. Reform of the Common Agricultural Policy (CAP) will have to precede the accession of major actual or potential food exporters such as Poland or Hungary. That is why the EU's eastern enlargement is turning into a protracted exercise. In other sectors, subsidies or market-sharing arrangements could be made, if the European Union was willing to accept further deviations from market-based resource allocation. Past evidence is that the European Union has usually been willing to accept such deviations, even if substantial net welfare-reducing trade diversion is incurred, in order to protect the 'sensitive sectors', but it has a cost and may be less acceptable in the future.

Bringing competing countries into ASEAN was much easier. Within Southeast Asia, intra-ASEAN trade has always been a small share of ASEAN members' total trade. Apart from flows between Singapore and Malaysia, and to a lesser extent Singapore and Indonesia, intra-ASEAN trade has been of minor importance (Pomfret, 1996). The ASEAN exporters of labour-intensive manufactures compete in Japanese, North American, and European markets, but each has a small total market share so rivalry is muted. One important element of ASEAN's evolution has been the recognition that there are benefits from negotiating as a group for market access. Although both effects are limited, enlargement added more to ASEAN's negotiating strength than to competition among ASEAN members for overseas markets.

The ASEAN countries also compete for inward foreign direct investment (FDI). Again, with respect to export-oriented FDI in labour-intensive activities, there is more sense of solidarity in setting conditions, than a sense of competing for limited FDI. This is less true when it comes to FDI catering for the Southeast Asian market. The car industry in particular has threatened to become a battleground, but the lines have been between national car projects in Malaysia and Indonesia and foreign-badged cars assembled in Thailand; none of the world's major automobile producers is likely to look to Indochina or Myanmar as a location for supplying the region. Thus, with respect to FDI as with trade, ASEAN enlargement has posed no major problem and, in the economic environment after the July 1997 enlargement, any competition for

FDI is further limited by the decline in capital inflows into the whole region.

In sum, although complementary partners seem to have most to gain and competing partners least to gain from a regional trading arrangement, the reality of the European Union and ASEAN may be different. Countries with large capacities in what the European Union has until now designated as 'sensitive products' pose a serious challenge to the whole edifice of a common market regulated by interventionist sectoral policies. Although the costs of price supports in agriculture and of subsidies and other interventionist measures in coal and steel have been shown to be large, they have so far not blown out the EU budget. A move from EU15 to EU25 with unchanged CAP would blow out the budget. Enlarging ASEAN to include the SEA-10, by contrast, has no major economic hitch. There are special time schedules for the new members to adopt the ASEAN Free Trade Area (AFTA), but AFTA is of minimal economic significance due to the limited intra-AFTA trade.[8]

CONCLUSIONS

Despite the similarity of the enlargement issues facing the EU and ASEAN during the 1990s the outcomes are likely to differ substantially. In the short-term ASEAN enlargement has proceeded smoothly (apart from a slight Cambodian hiccup) whereas EU enlargement has not proceeded at all (apart from the German adjustment). Enlargement will consume EU political attention, but will have no economic impact this century. In ASEAN there will be some disputes over intra-ASEAN trade as new members have limited experience with WTO-type obligations and seek exception status in implementing AFTA, but because intra-ASEAN trade will continue to be of minor importance internal trade disputes are unlikely to be major. ASEAN will assume slightly greater weight in international negotiating now that it consists of nine rather than six countries and its total population has increased from 328 million in 1992 to over 470 million in 1997. The larger group will, moreover, remain fairly cohesive in its negotiating aims as, with the exception of the two small states, Brunei and Singapore, the ASEAN countries all want the high income countries to have open markets for labour-intensive manufactures and food and a few other primary products. In the longer term the outcome is less predictable. Some at least of the Eastern European countries are likely to become EU members early in the next century. This will increase EU self-sufficiency in some ASEAN export items and perhaps involve some trade diversion at ASEAN countries' expense. On the other hand, the prospects for enlargement behind high external trade barriers are slim, especially now that agriculture has been reincorporated into the WTO framework for international trade.

In sum, the prospects are positive for increasing EU–ASEAN economic relations. The balance of economic power will clearly favour the EU for the foreseeable future, but the enlargement process is likely to strengthen ASEAN's relative negotiating strength. Since that strength will be used to reinforce liberal tendencies within the EU, there should be a positive outcome for all apart from some protected producers in the EU15.

NOTES

1. ASEAN was founded in 1967 by Indonesia, Malaysia, the Philippines, Singapore, and Thailand. Brunei joined in 1984, but with a population of only a third of a million and high per capita income its accession required little adjustment. Vietnam joined in July 1995, and Laos and Myanmar in July 1997, when Cambodia was also scheduled to become a member but its accession was postponed. SEA-10 refers to all of the above countries, and is widely believed to represent ASEAN's geographical limits as a Southeast Asian organization. Pomfret (1996) reviews ASEAN's evolution and prospects.
2. Norway, Switzerland, Iceland, Malta, and Cyprus remained outside, although they were linked to the European Union by varying arrangements.
3. Geographically expansion to the south-east is conceivable to include Papua New Guinea (PNG), Australia, and New Zealand, but appeared to be considered as a cultural leap. In the 1990s, PNG did sign the Treaty of Amity and Cooperation, often considered the first step to ASEAN membership, and Australia and New Zealand held tentative talks with ASEAN about linking it to the Closer Economic Relations agreement in a regional arrangement.
4. The first group of applicants, embarking on formal negotiations on 31 March 1998, consisted of Poland, Hungary, the Czech Republic, Estonia, and Slovenia.

5. Tangerman (1997) reports EU Commission lower-bound estimates that expenditure under an unchanged CAP would rise by 12 billion ecu (a 28 per cent increase) if the 10 Eastern European countries joined. That is well above internal EU limits on spending growth, and any raising of the limits would be unpopular among Western European taxpayers and contrary to the EU's WTO commitments for phasing out agricultural subsidies. See also Baldwin *et al.* (1997) on EU enlargement.

6. This is not to deny that reform has been glacially slow in some areas, most notably reforming State-owned enterprises which have changed little and remain more like Western nationalized industries than private enterprises.

7. Agriculture also played other important roles in providing markets, capital, and exports. By sharing the benefits with the rural population, an agriculture-led strategy co-opted a large part of the population in support of the reform process.

8. AFTA calls for tariffs of 5 per cent or less on intra-ASEAN trade, but does not restrict external trade policies. With free-trading Singapore as a potential entrepôt, a true free trade area would need to address the trade deflection problem (Pomfret, 1997a: 185–8), but even 5 per cent tariffs which require continuation of customs procedures could be sufficient to deter transshipment.

BIBLIOGRAPHY

Baldwin, R., Francois, J. and Portes, R. (1997) 'The Costs and Benefits of Eastern Enlargement', *Economic Policy*, 24, 125–70.

Blanchard, O. (1997) *The Economics of Post-Communist Transition* (Oxford: Clarendon Press).

Chen, Kang, Jefferson, G. and Singh, I. (1992) 'Lessons from China's Economic Reform', *Journal of Comparative Economics*, 16, 201–25.

Fischer, S., Sahay, R. and Végh, C. (1997) 'How Far is Eastern Europe from Brussels?' In H. Siebert (ed.), *Quo Vadis Europe?* (Tübingen: JCB Mohr).

Langhammer, R. (1997a) 'How Far is Indochina from ASEAN?' *ASEAN Economic Bulletin*, 14, November, 160–76.

Langhammer, R. (1997b) 'EU Enlargement: Lessons for ASEAN'. Paper presented at the Institute of Southeast Asian Studies, Singapore, November 20–21.

Pomfret, R. (1996) 'ASEAN: Always at the Crossroads?' *Journal of the Asia Pacific Economy*, 1, 365–90.

Pomfret, R. (1997a) *The Economics of Regional Trading Arrangements.* (Oxford: Clarendon Press).

Pomfret, R. (1997b) 'Growth and Transition: Why has China's Performance been so different?' *Journal of Comparative Economics*, 25, 422–40.

Sachs, J. and Wing Thye Woo (1994) 'Structural Factors in the Economic Reforms of China, Eastern Europe, and the Former Soviet Union', *Economic Policy*, 9, 101–45.

Tangerman, S. (1997) 'Reforming the CAP: A Prerequisite for an Eastern Enlargement'. In H. Siebert (ed.), *Quo Vadis Europe?*, pp. 151–79 (Tübingen: JCB Mohr).

Conclusions

Jim Slater

Roger Strange in his Introduction to this volume referred to the April 1998 Conference in Como, Italy at which many of the chapters were initially presented. Some of the papers had been in preparation for a year or two and made little or no reference to the Asian crisis. A smaller number addressed the crisis head-on. Opinion on the severity of the crisis was largely polarized between authors of the two types of contribution. With some exceptions, researchers who had been motivated to study the high performing economies argued that the evident major weaknesses in financial systems would have minimal impacts on the real economies and recovery could be expected in the short-to-medium term. The second group were of the opinion that the real economies were shallow-rooted and that the financial crisis would cut deep. Regarding the central question, namely, the state of EU–ASEAN economic relations, the analyses point in similar directions: trade and investment links are not strong and have in relative terms been declining. Furthermore, relations are asymmetric with the EU being more important to the ASEAN countries than ASEAN is to the EU. It seems reasonable to try to disentangle the two main viewpoints by decoupling trends in the relationship and related policy issues from the dislocation arising from the crisis.

Inward investment into and outward investment from ASEAN have undoubtedly grown rapidly in the last decade. So has trade. The greatest contribution, by far, to the internationalization process has originated from within the East Asian region as a whole. Aggregate inward investment has without doubt contributed (especially since the last crisis around the mid-1980s) significantly to rapid domestic growth, but there are qualitative dimensions to the flows of funds and differences in the various ASEAN countries which warn against over-generalization. For example Singapore, in contrast to the other ASEAN economies, has sustained balance of trade surpluses, has attracted a high proportion of technologically advanced investment from multinationals headquartered in advanced nations, and has pursued (until recently) a policy of insulating offshore financial activities from the domestic economy. Even so, Singapore's major markets are still regional and vulnerable to local shocks. At the other extreme of economic development the new ASEAN member states have relatively little economic superstructure and are, therefore, less susceptible to damage from financial turbulence. In between, in economic development terms, Thailand, Malaysia, the Philippines, and Indonesia differ in culture, financial institutions, accounting procedures (including issues of transparency and governance), demographics, political and industrial structure, and composition of trade and investment flows. Some two years from the onset of the Asian crisis, it is perhaps unsurprising that the ASEAN economies have absorbed the impact differentially.

Links with Europe have weakened relatively in the post-colonial period, but within the loosened ties the old European partners are still well represented. It is unlikely that initiatives such as ASEM will affect macro-aggregates other than on the margin, but at micro-levels there is scope for mutual benefit.

The impact of the Asian dislocation on the EU depended (and may still depend) upon the resilience of the international financial system as a whole. Predictions of competitive devaluation involving China, leading to destabilization of Japan's stagnant economy and parlous banking system have yet to be realized. The direct links between ASEAN and the EU, as evaluated in the chapters in this book, are insufficiently strong on both sides to give rise to dependency. Therefore, the direct effects of recession in the ASEAN countries on the EU were always likely to be small. The indirect effects of a chain reaction, feared in Europe as elsewhere, through Japan to the US economy, potentially leading to global recession, have not to date materialized.

Regarding economic recovery in the ASEAN countries, first and second quarter indicators for 1999, where available, show upward trends in inward investment, output, and surpluses in balances of trade. Although these indicators are short term, they suggest that the worst has been suffered in Singapore, Malaysia, and Thailand (South Korean figures are positive, too). Indonesia is a different case, having been hardest hit in exchange rate terms fuelling social unrest, capital destruction, and flight of expertise as well as capital. Recent Indonesian balance of trade surpluses reflect falls in imports rather than export strengths. Although recent elections may help restore confidence, long-term restoration of the economic and business infrastructure will not be realized overnight. In some of the other countries there are already warning signs that some of the on-going problems in the financial sector have not been resolved. For example, stock market price indices in Thailand and Malaysia have risen rapidly, raising fears of volatility once controls are relaxed.

It would be trite, but probably true, to conclude that current economic relations between the ASEAN states and the EU understate the potential for mutual gains from both trade and investment. Government policies in most ASEAN countries underline the importance of technology transfer, for example, and there is also clear scope for infrastructural development. There are still considerable ASEAN funds available for investment (much was moved offshore by knowledgeable insiders prior to the crash). Prudent European, and other external, investors are, however, likely to exercise caution in the near future. The crisis has been a reminder that enthusiastic speculation in poorly regulated and opaque financial systems inevitably leads to bursting bubbles. This is hardly outside the European experience, but speculation, by its nature, is not subject to learning by experience. ASEAN financial systems must evolve to reduce the incentives for the wilder excesses. This is easily observed, but implementation is a different matter when the roots of this type of behaviour are deeply institutionally and culturally embedded. Even though the recent dislocation has exposed these systemic problems there have been and will continue to be questions of political risk. Most of the ASEAN countries have experienced phenomenal growth rates under strong, pragmatic leaders who have been in power for considerable lengths of time. Even in Thailand, where governments have changed frequently, stability has been maintained through the population's trust and reverence for the King. Whatever the debates surrounding the current strong men, the uncertainty of political succession in countries with deep political tensions remains.

Index

accommodation, tourist 171–2, 174–5
acquisitions and mergers 79, 80, 131
agglomeration effects 212
agricultural labour 228
agriculture 182
  and enlargement of EU 233, 234
  exports from ASEAN 18
Akamatsu, K. 14
Albania 111
anti-dumping (AD) measures 16–17
Apoteker, T. 203
Arthur Andersen surveys 67–9, 79
ASEAN Free Trade Area (AFTA) 29, 64, 77, 80, 82, 200, 234
ASEAN Investment Area (AIA) 64–5, 77, 82
  provisions 80–1
ASEM Connect 94
Asia–Europe Meetings (ASEM) 1, 14–15, 30, 87, 156
  Trust Fund 135–6
Asia–Pacific Economic Cooperation (APEC) 9, 14, 200
Asia–Pacific region 9
Asian developing countries 66–8, 179
Asian Development Bank 150, 157
Asian financial crisis 1, 3, 82, 117–39, 140–58
  assessment of 140–7; capital inflow and credit inflation 144–5; speculation in real estate 145–6
  description and explanation 121–6
  and EU–ASEAN trade 27–9, 127–31

flaws in Asian model of industrialization 147–50
future prospects for Asia 154–7
IMF rescue packages 150–2; criticisms of conditions attached 152–4
impact on EU 127–34, 156–7; finance 131–4; trade and investment 127–31
long-term effects on ASEAN–EU relations 238–40
missed opportunity for EU 134–6
theory of currency crises 118–20
TNCs' investment strategies after 78–81
Asian NIEs
  FDI in ASEAN 47–50; regionalism and 51–5
  FDI in Vietnam 202, 204
Association of Southeast Asian Nations (ASEAN) 1, 5, 235
  Cooperation Agreement with EU 14
  and the 'East Asian Miracle' 10–14
  enlargement *see* enlargement
  FDI *see* foreign direct investment; outward direct investment
  future prospects for ASEAN–EU relations 156–7, 238–40
  industrialization strategies 34–44; colonial era 34–5; post-independence era 35–44
  tourism 162–4
  trade with EU *see* trade
  Vietnam's economic reforms and accession to 200–1
asymmetric shocks 130
auction of loans and assets 154–5

241